MEMOIRS
of the
American Mathematical Society

Number 454

Historical Processes

Donald A. Dawson
Edwin A. Perkins

September 1991 • Volume 93 • Number 454 (end of volume) • ISSN 0065-9266

American Mathematical Society
Providence, Rhode Island

1980 *Mathematics Subject Classification* (1985 *Revision*).
Primary 60H15, 60J80, 60J60.

Library of Congress Cataloging-in-Publication Data

Dawson, Donald A., 1937–
Historical processes/Donald A. Dawson, Edwin A. Perkins.
p. cm. – (Memoirs of the American Mathematical Society, ISSN 0065-9266; no. 454)
Includes bibliographical references.
ISBN 0-8218-2508-9
1. Stochastic partial differential equations. 2. Branching processes. 3. Diffusion processes.
I. Perkins, Edwin Arend, 1953– . II. American Mathematical Society. III. Title. IV. Series.
QA3.A57 no. 454
[QA274.25]
510 s–dc20 91-22744
[519.2] CIP

Subscriptions and orders for publications of the American Mathematical Society should be addressed to American Mathematical Society, Box 1571, Annex Station, Providence, RI 02901-1571. *All orders must be accompanied by payment.* Other correspondence should be addressed to Box 6248, Providence, RI 02940-6248.

SUBSCRIPTION INFORMATION. The 1991 subscription begins with Number 438 and consists of six mailings, each containing one or more numbers. Subscription prices for 1991 are $270 list, $216 institutional member. A late charge of 10% of the subscription price will be imposed on orders received from nonmembers after January 1 of the subscription year. Subscribers outside the United States and India must pay a postage surcharge of $25; subscribers in India must pay a postage surcharge of $43. Expedited delivery to destinations in North America $30; elsewhere $82. Each number may be ordered separately; *please specify number* when ordering an individual number. For prices and titles of recently released numbers, see the New Publications sections of the NOTICES of the American Mathematical Society.

BACK NUMBER INFORMATION. For back issues see the AMS Catalogue of Publications.

MEMOIRS of the American Mathematical Society (ISSN 0065-9266) is published bimonthly (each volume consisting usually of more than one number) by the American Mathematical Society at 201 Charles Street, Providence, Rhode Island 02904-2213. Second Class postage paid at Providence, Rhode Island 02940-6248. Postmaster: Send address changes to Memoirs of the American Mathematical Society, American Mathematical Society, Box 6248, Providence, RI 02940-6248.

10 9 8 7 6 5 4 3 2 1 95 94 93 92 91

CONTENTS

ABSTRACT

The historical process is constructed to be a superprocess which is enriched so as to contain information on genealogy. This process together with a representation for the associated Palm measures are used to describe the modulus of continuity and equilibrium structure for a class of superprocesses, and to establish that the super-Brownian motion in dimensions $d \geq 3$ has constant density with respect to the appropriate Hausdorff measure.

Received by editor February 12, 1990.

KEY WORDS AND PHRASES: Superprocess, historical process, Palm measure, Hausdorff measure, modulus of continuity, equilibrium measure.

1. Introduction.

Consider a Markov process Y taking values in a Polish space, E. For $M \in \mathbb{N}$, $K(M)$ particles start at $x_1^M(\omega), \ldots, x_{K(M)}^M(\omega) \in E$ and follow independent copies of Y on the time interval $[0,1/M)$. At $t = 1/M$ each particle dies or splits into two, each with probability 1/2, independently of each other. The new individuals then follow independent copies of Y on $[1/M, 2/M)$, and this pattern of alternating spatial motions and branching mechanisms continues indefinitely. In this way we construct a random "Y-tree" of branching particles. Define a random measure on $(E, \mathcal{E})$ ($\mathcal{E}$ is the Borel σ-field) by

$$X_t^M(\omega)(A) = M^{-1}(\text{no. of particles in A at time t}), \ A \in \mathcal{E}, \ t \geq 0.$$

X^M is a process taking values in the space $M_F(E)$ of finite measures on E with the topology of weak convergence. Watanabe (1968) (see Ethier-Kurtz (1986, Ch. 9) for the required tightness) showed that if Y is a Feller process and E is locally compact, then $X_0^M \longrightarrow m_0$ (in $M_F(E)$) implies $X^M \xrightarrow{w} X$ on $D([0,\infty), M_F(E))$ (weak convergence of laws), where X is a $M_F(E)$-valued diffusion starting at m_0. Let Q_{m_0} denote the limiting law on the appropriate space

---

Research of both authors partially supported by the Natural Sciences and Engineering Research Council of Canada.

of $M_F(E)$-valued paths. The limit X is the Y-superprocess (or DW-superprocess) in the terminology of Dynkin. Watanabe first constructed X by specifying its semigroup through its Laplace functional $Q_{m_0}(\exp\{-\langle X_t,\phi\rangle\})$, $\phi\geq 0$ ($\langle\nu,\phi\rangle = \int\phi(x)d\nu(x)$), and this approach has recently been extended by Dynkin (1989b,d) and Fitzsimmons (1988) to a much more general class of processes Y. In Section 7 we prove Watanabe's weak convergence result when Y is a Hunt process taking values in a Polish space (Theorem 7.13). The proof relies on a general martingale problem for X due to Fitzsimmons (1988,1989). (See Dynkin (1989d) for a proof giving the convergence of finite dimensional distributions in a more general setting.)

To motivate the purpose of this work consider the following elementary problem: If $S(\nu)$ denotes the closed support of a measure ν, Y is continuous and $S(X_0)$ is connected, is $R = \bigcup_{t\geq 0} S(X_t)$ (the "range" of X) connected? To show that it is, simply trace a particle in $S(X_t)$ back to time 0 through its ancestors, along what should be a continuous trajectory of Y. This allows us to connect any point in R to $S(X_0)$ and since $S(X_0)$ is connected we are done. There are two technical problems here. The total length of the Y-tree approaches ∞ as $M\to\infty$ and so the continuity of each branch of the tree is not assured (in fact without an additional condition on Y it may fail - see Example 8.16). One must check that the past history of each particle is contained in R since "thin spots" could develop in the limiting operation. If one is working with Q_{m_0} on canonical path space, however, there is a much more serious problem; it does not seem possible to recover the ancestry of a given individual from X. Apparently important information has been lost in the limit as $M\to\infty$. For super-Brownian motion (i.e. Y is a Brownian motion on $\mathbb{R}^d$) a nonstandard model which contains the genealogical trees of all the points in R was introduced in Perkins (1988) and the aforementioned technical problems were solved for this model in Dawson, Iscoe and Perkins (1989) (hereafter abbreviated [DIP]) (Lemmas 4.7, 4.9). Our main goal is to introduce a standard process, the historical process, which records the past histories of all individuals in the population and show how this process may be used to study properties of superprocesses.

To describe the historical process, consider the Y-tree of branching particles. Let H_t^M be the random measure on the space D = D(E) of E-valued càdlàg paths on $[0,\infty)$ (with the J_1-topology) which assigns mass M^{-1} to the trajectory of each particle alive at time t (the trajectories are stopped at time t). If Y is a Hunt process we show in Section 7 (Theorem 7.15) that H^M converges weakly in $D(M_F(D))$ to an $M_F(D)$-valued diffusion, H_t, the "Y-historical process". In fact this result is an easy consequence of the weak convergence result (Theorem 7.13) already described. We need only replace the E-valued Hunt process Y with the time-inhomogeneous D(E)-valued Hunt process $W_t = Y(.\wedge t)$ and apply Theorem 7.13 to the appropriate space-time process. Evidently H is just the W-superprocess and this observation allows us to construct H directly from the general results of Dynkin and especially Fitzsimmons (1988). Although Dynkin's results are well-suited because of their time-inhomogeneous setting we will for the most part use Fitzsimmons (1988) (hereafter abbreviated

[F]) because of the additional regularity results he obtained on the paths of X.

We now give a more precise description of the setting for most of this work.

Assume

(M_1) $Y = (D, \mathcal{D}, \mathcal{D}_{t+}, \theta_t, Y_t, P^x)$ is a Borel right process with càdlàg paths in the Polish space E.

Here we are working with the canonical representation of Y on path space and $\mathcal{D}_{t+}$ is the canonical right continuous filtration on D = D(E). A continuous state branching mechanism (Kawazu and Watanabe (1971)) is given by

(1.1) $$\Phi(x,\lambda) = -b(x)\lambda - c(x)\lambda^2/2 + \int_0^\infty (1-e^{-\lambda u}-\lambda u)n(x,du), \quad x \in E, \quad \lambda \ge 0,$$

where

(1.2) b,c are bounded measurable functions from E to $\mathbb{R}$ and $[0,\infty)$, respectively, and n(x,du) is a measurable kernel from E to $\mathcal{B}(0,\infty)$ such that $\sup_x \int (u \wedge u^2) n(x,du) < \infty$.

Fitzsimmons (1988, 1989) shows the existence of a (Y,Φ)-superprocess, X. X is a Borel right process with càdlàg paths in $M_F(E)$, and is a Hunt process if Y is. See Theorem 2.1.3 for a detailed description of this result and the semigroup of X. The superprocess constructed above as a weak limit of a branching particle system is the $(Y,-\lambda^2/2)$-superprocess (or simply Y-superprocess). We will often consider the particular branching system

(SB) $$\Phi(x,\lambda) = \Phi(\lambda) = -\gamma\lambda^{1+\beta}, \quad 0 < \beta \le 1, \ \gamma > 0.$$

In the particle picture the offspring distributions are now particular mean one laws in the domain of attraction of a one-sided $(1+\beta)$-stable law (finite variance equal to 2γ if $\beta=1$) and each particle is assigned mass $M^{-1/\beta}$ (see Méléard and Roelly-Coppoletta (1989) for a proof in the Feller setting).

If $y \in D$, let $y^t(u) = y(u \wedge t)$ and let $D^t = \{y^t : y \in D\}$. It is not hard to check that $W_t = Y^t \in D^t$, which we refer to as the path process, is a time-inhomogeneous Borel strong Markov process with càdlàg paths in the Polish space D, and is also a Hunt process if Y is (see the beginning of Section 2 for relevant definitions and Theorem 2.2.1 for a proof.) The (Y,Φ)-historical process is the (W,Φ)-superprocess, where Fitzsimmons' construction is extended to the time inhomogeneous setting by working with the W-space-time process as usual (the tedious details are given in Section 2). Hence the (Y,Φ)-historical process H_t is an inhomogeneous Borel strong Markov process with càdlàg paths in $M_F(D)$. An explicit description of its semigroup and additional regularity properties of H may be found in Theorem 2.2.3. Keeping the particle picture in mind, it seems obvious that $X_t = H_t \circ \Pi_t^{-1}$ ($\Pi_t(y) = y(t)$ is the coordinate mapping on D) is the (Y,Φ)-superprocess. This is verified in Theorem 2.2.4.

The existence of H required only the Markov property of W, but not that of Y. Section 2.1 deals with this more abstract framework in which no Markovian assumptions are placed on Y, and Φ may depend on paths in D (see Theorems 2.1.5 and 2.1.7 for a description of H in this setting). This has the potential of leading to a "general theory of superprocesses". The Markovian setting described above is treated in Section 2.2 and is the setting for the rest of the paper.

We assume (SB) throughout the rest of this Introduction.

Section 3 gives a probabilistic a.s. construction of H_t. Keeping the particle picture in mind, it is easy to see that $H_t|_{\mathcal{D}_{t-\varepsilon}}$ is supported on a finite number of atoms. In fact, conditioned on $H_{t-\varepsilon}$, the number of atoms has a Poisson law with mean $H_{t-\varepsilon}(D)(\beta\gamma\varepsilon)^{-1/\beta}$ (see Proposition 3.3). We identify the branching rate and offspring distribution of the system of trajectories obtained by laying down the atoms in $H_t|_{\mathcal{D}_{t-\varepsilon}}$ at time t and letting t vary (Theorem 3.9). We then show that if we use the law of Y on $[t-\varepsilon,t]$ to extend each atom of $H_t|_{\mathcal{D}_{t-\varepsilon}}$ to $\mathcal{D}_t$ and give each such contribution total mass $(\varepsilon\beta\gamma)^{1/\beta}$, then the resulting random measures (on $\mathcal{D}_t$) converge a.s., as $\varepsilon \downarrow 0$, to H_t for fixed t (Theorem 3.10). These results are closely tied to a recent paper of Le Jan (1989) and are independent of the rest of this paper.

The description of the semigroup of H in Theorem 2.2.3 shows H is an infinitely divisible random measure on D (this is also obvious from the particle picture). At the beginning of Section 3 we recall the "Lévy-Khinchine" representation of H_t in terms of its canonical (or Lévy) measure $R_{s,t}(y,d\nu)$ $(y\in D)$. $R_{s,t}(y,d\nu)$ is a finite measure on $M_F(D)$ of total mass $2/(t-s)$ where for the sake of this discussion we are setting $\beta=1$ and $\gamma = 1/2$. A simple probabilistic interpretation of $R_{s,t}(y,d\nu)$ may be given in terms of the particle picture. If $y\in D^s$ is the path of a particle up to time s we call the collection of particles at time t which are descendents of y, a y-clan (or simply a clan if s=0). $R_{s,t}(y,d\nu)/(2/(t-s))$ is the conditional law on $M_F(D)$ of the past histories of the y-clan at time t, given that this clan is non-empty. The total mass $R_{s,t}(M_F(D)) = 2(t-s)^{-1}$ gives the rate at which clans are formed. The Palm and Campbell measures of H_t and the canonical measures $R_{s,t}(y,.)$ are introduced in Section 4.1. The Palm measure $(R_{s,t})_y(d\nu)$ of the canonical measure is the law of the y^s-clan (at time t) which contains the "tagged particle" $y \in D^t$. Here y is a typical path in $S(H_t)$ chosen according to H_t. $(R_{0,t})_y(d\nu)$ is clearly composed of independent clusters which break off from y on [0,t]. A cluster will break off from y at time s with rate 2/(t-s) and if it does, it should have law $R_{s,t}(y^s,d\nu)/(2/(t-s))$. Hence $(R_{0,t})_y(d\nu)$ should be the law of $\int_0^t \nu N(ds,d\nu)$ where N is a Poisson point process on $[0,t]\times M_F(D)$ with characteristic measure

(1.3) $\mathbf{1}(s\le t)R_{s,t}(y^s,d\nu)ds.$

This is proved in Section (4.1) (Proposition 4.1.5) by an analytic argument. The key step is a simple Feynman-Kac argument (Theorem 4.1.3) which also shows that this interpretation of the Palm measure is valid if $\beta<1$ and ds in (1.3) is replaced by "$d\tau(s)$" where τ is a stable subordinator of index β. This representation is used in Section 4.2 to prove a "0-1" law for the Campbell measure of H_t (Theorem 4.2.2) when Y is a Lévy process in $\mathbb{R}^d$. This 0-1 law states that events, which depend on the relative positions, y_t and $\tilde{y}_t$, of a typical y in $S(H_t)$ and arbitrarily closely-related cousins $\tilde{y}$, are trivial events. (This result was first established for the nonstandard model.) The above representation theorem reduces the proof to a simple application of the Kolmogorov and Blumenthal 0-1 laws.

In Section 5 the above 0-1 law is used to refine results in Perkins (1988, 1989) on the Hausdorff measure of the supports of super-symmetric stable processes. Let Y_α denote a symmetric stable process of index α on $\mathbb{R}^d$ scaled so that

$$P^0(e^{i(\theta,Y_\alpha(t))}) = \begin{cases} \exp\{-|\theta|^\alpha t\} & \text{if } \alpha<2 \\ \exp\{-|\theta|^2 t/2\} & \text{if } \alpha=2 \end{cases}.$$

Let $\psi_\alpha(r) = r^\alpha \log\log \frac{1}{r}$ and denote ψ_α-Hausdorff measure by ψ_α-m. Let X denote the $(Y_\alpha,-\lambda^2/2)$-superprocess. If $d>\alpha$, then (Perkins (1988)) there are $0 < c(\alpha,d) \le C(\alpha,d) < \infty$ such that for a.a. ω and all $t > 0$ there is a Borel set $\Lambda_t(\omega)$ supporting $X_t(\omega)$ such that

(1.4) $\quad c(\alpha,d)\psi_\alpha\text{-m}(A\cap\Lambda_t) \le X_t(A) \le C(\alpha,d)\psi_\alpha\text{-m}(A\cap\Lambda_t)$ for all $A\in\mathcal{B}(\mathbb{R}^d)$.

If $\alpha=2$ (i.e. X is super-Brownian motion), then (1.4) holds with $S(X_t)$ in place of Λ_t (Perkins (1989)), and this extension is false for $\alpha < 2$ (Perkins (1990)). In Section 5 we use the 0-1 law to show that the Radon-Nikodym derivative of X_t with respect to ψ_α-m$(\cdot\cap\Lambda_t)$ (and with respect to ψ_α-m$(\cdot\cap S(X_t))$ if $\alpha=2$) is a.s. constant and hence are able to conclude that $c(\alpha,d)=C(\alpha,d)$ in the above results at least for t fixed (see Theorems 5.2 and 5.5). An immediate consequence in the Brownian case is the Markov property of the set-valued process $S(X_t)$.

In Section 6 we study the equilibrium (or stationary) distributions of X and H under (SB) and when $Y=Y_\alpha$. It is known (Dawson (1977), Dynkin (1989c), Gorostiza, Roelly-Coppoletta and Wakolbinger (1989)) that if $d > \alpha/\beta$ and $X_0 = dx$, then X_t converges weakly to a stationary distribution as $t \to \infty$. We give an elementary proof of this using our representation for the Palm measure $(R_{s,t})_y$ from Section 4 (Proposition 6.1). By the stationarity of Y_α (with respect to Lebesgue measure) we may start X at time $-t$ with Lebesgue measure and observe X_0 and H_0 as $-t \longrightarrow -\infty$. In this case the Palm $(R_{-t,0})_Y$ at time 0 is a Poisson superposition of clans branching out from $[-t,0]$ and as $-t \longrightarrow -\infty$ these Palm measures converge by monotonicity. The transience condition $d > \alpha/\beta$ is used to show the total mass of bounded sets remains bounded as $-t \longrightarrow -\infty$. The convergence of the Palm measures and the fact that the mean measure is constant (Lebesgue measure) implies the weak convergence of X_0 as $-t \longrightarrow -\infty$ to an equilibrium random measure X_∞. A similar argument shows that under the same conditions H_0 converges weakly to an equilibrium distribution, H_∞, on $D((-\infty,0])$ (Theorem 6.3). A uniform integrability argument shows that $X_\infty = H_\infty \circ \Pi_0^{-1}$ has Lebesgue mean measure thus establishing the property known as persistence. The infinitely divisible equilibrium distributions X_∞ and H_∞ are studied through their canonical measures which govern the contributions of unrelated clans to these random distributions. In particular these canonical measures and the associated Palm measures are shown to be self-similar (Theorems 6.6, 6.7). The latter allows us to apply a result of Zähle (1988) to show that for a fixed $t > 0$, X_t may be supported on a Borel set of Hausdorff dimension α/β ($<d$) but does not charge any set of smaller dimension (Theorem 6.9).

Section 7 contains the weak convergence results already described. These results are then used in Section 8 to study path properties of $S(X_t)$ and $S(H_t)$ when $\Phi(x,\lambda) = -\lambda^2/2$ and Y is a diffusion taking values in the Polish space E. A mild regularity condition on Y (see (8.1), (8.2) and Remark (8.2(c)) implies that

$S(H_t)$ and $S(X_t)$ are compact sets in $C([0,\infty),E)$ and E, respectively, for all $t > 0$ a.s. (Theorem 8.10). In addition a uniform modulus of continuity is found for all the paths in $\bigcup_{t>0} S(H_t)$ (Theorem 8.6). Intuitively, all the branches of the Y-tree remain continuous as $M \to \infty$. If Y is a Brownian motion in $\mathbb{R}^d$ these results imply that for $c > 2$ there is a $\delta(c,\omega) > 0$ a.s. such that for any $t > 0$ and $y \in S(H_t)$ if $0 < t-s < \delta(c,\omega)$, then $|y_t - y_s| < c[(t-s)\log(1/(t-s))]^{1/2}$. We also show this conclusion fails a.s. for $c < 2$ (Theorem 8.7). This particular result was proved for the nonstandard model for H in [DIP, Thm. 4.7] and the arguments given here in the more general setting are similar. It is interesting to note that these results may fail rather badly even for a 2-dimensional diffusion if no additional regularity condition is imposed on Y (see Example 8.16). Finally the elementary connectedness problem described at the beginning of this section is solved (in the affirmative) under the same regularity condition on Y (see Theorem 8.15 and also Le Gall (1989c, Sec. 5)).

Enriched models containing additional information on the family trees of individuals have been studied in various models in population genetics (Donnelly and Tavare (1987), Donnelly and Kurtz (1989)). The role of the family structure of continuous time branching particle systems has also received considerable attention (Gorostiza (1981), Neveu (1986), Chauvin (1986), Le Jan (1989), Gorostiza and Wakolbinger (1989)). Earlier these considerations had played a role in the study of discrete time processes (Fleischmann and Prehn (1974), Fleischmann and Sigmund-Schultze (1977)) and in particular in the backward tree formula of Kallenberg (1977a) in his criterion for persistence. Even in the context of superprocesses, the historical process has also been independently introduced by Le Gall (1989b,c) and Dynkin (1989d). Our construction has the advantage of using the existing machinery of superprocesses. The applications we present here are for the most part distinct from those in these two papers although in Sections 8 and 4.1 there is some overlap with Le Gall (1989c).

We close this section with some notation. If $\mathcal{F}$ is a σ-field $b\mathcal{F}$ (respectively, $p\mathcal{F}$) denotes the set of bounded (respectively, positive) $\mathcal{F}$-measurable functions and $bp\mathcal{F} = b\mathcal{F} \cap p\mathcal{F}$.

If E and S are topological spaces let

$\mathcal{B}(E)$ = Borel σ-field on E,

$M_1(E)$ = set of probabilities on $E \subset M_F(E)$,

$C(S,E) = \{f:S\to E,\ f \text{ continuous}\}$ with the compact open topology,

$C(E) = C([0,\infty),E)$, $C_b(E) = C(E,\mathbb{R})\cap b\mathcal{B}(E)$.

If $P_t: b\mathcal{E} \longrightarrow b\mathcal{E}$ and $m\in M_F(E)$ let $mP_tf = \langle m,P_tf\rangle$.

If X is a random variable, $E(X)$ or $P(X)$ denotes its expectation.

$C_c^\infty(\mathbb{R}^d)$ is the class of infinitely differentiable functions with compact support on $\mathbb{R}^d$.

[DM] denotes Dellacherie and Meyer (1978).

Acknowledgement. We thank P. Fitzsimmons for answering many queries on his work, E.B. Dynkin for keeping us informed of his work in this area, I. Iscoe for discussions on the formulation of historical processes, J.F. Le Gall, L.G. Gorostiza, S. Roelly-Coppoletta and A. Wakolbinger for discussions concerning the Palm measure formula, K. Fleischmann for discussions on the reduced tree method and T. Kurtz for discussion concerning Lemma A.5.

2. Definitions, Preliminaries and Generalities.

2.1 The General Setting.

If $y \in D(S)$ let $y^t(s) = y(s \wedge t)$.

Definition 2.1.0 Let $(S,\mathcal{S})$ be a metrizable Lusin space (i.e. S is homeomorphic to a Borel subset of a compact metric space) with its Borel σ-field. Let $\hat{E} \in \mathcal{B}([0,\infty)) \times \mathcal{S}$ and set $S^t = \{x:(t,x) \in \hat{E}\}$. We say $Z = (\Omega,\mathcal{F}^o,\mathcal{F}^o[s,t+],Z_t,P_{s,z})$ is an inhomogeneous Borel strong Markov process (IBSMP) with cadlag paths in $S^t \subset S$ iff:

(i) $(\Omega,\mathcal{F}^o)$ is a measurable space and $\{\mathcal{F}^o[s,t]:s \le t\}$ is a non-decreasing collection of sub-σ-fields of $\mathcal{F}^o$ indexed by compact intervals. $\mathcal{F}^o[s,t+] = \bigcap_{n=1}^{\infty} \mathcal{F}^o[s,t+1/n]$, $\mathcal{F}^o[s,\infty) = \bigvee_n \mathcal{F}^o[s,n]$.

(ii) $\forall\ (s,z) \in \hat{E}$, $P_{s,z}$ is a probability on $(\Omega,\mathcal{F}^o[s,\infty))$ such that $\forall A \in \mathcal{F}^o[u,\infty)$, $(s,z) \longrightarrow P_{s,z}(A)$ is Borel measurable on $\hat{E} \cap ([0,u] \times S)$.

(iii) $\forall\ t \ge 0$, $Z_t:(\Omega,\mathcal{F}^o[t,t]) \longrightarrow (S,\mathcal{S})$ is measurable and satisfies

$P_{s,z}(Z(s)=z,\ Z_t \in S^t\ \forall t \ge s$ and $Z_{.}$ is càdlàg on $[s,\infty)) = 1$ $\forall (s,z) \in \hat{E}$.

(iv) If $(s,z) \in \hat{E}$, $\psi \in b\mathcal{B}([s,\infty) \times D(S))$ and $T \ge s$ is a stopping time with respect to $\{\mathcal{F}^o[s,t+]:t \ge s\}$, then

$P_{s,z}(\psi(T,Z(T+.))\,|\,\mathcal{F}^o[s,T+])(\omega) = P_{T(\omega),Z(T)(\omega)}(\psi(T(\omega),Z(T(\omega)+.)))$, $P_{s,z}$-a.s. on $\{T<\infty\}$.

We say that Z is an inhomogeneous Hunt process if, in addition,

(v) $\forall (s,z) \in \hat{E}$, if $T_n \ge s$ are $\{\mathcal{F}^o[s,t+]:t \ge s\}$-stopping times which increase to T and $P_{s,z}(T<\infty) = 1$, then $Z(T_n) \longrightarrow Z(T)$ $P_{s,z}$-a.s.

It is easy to check that (ii) and (iii) imply that

$\forall\psi\in\mathcal{B}(D(S))$,

(iii)' $(s,z,t) \longrightarrow P_{s,z}(\psi(Z(t+.)))$ is Borel measurable on $\{(s,z,t)\in\hat{E}\times[0,\infty):t\geq s\}$ and hence the inhomogeneous semigroup $T_{s,t}\phi(z) = P_{s,z}(\phi(Z_t))$ is jointly measurable in (s,z,t) on the same set. Note that $T_{s,t}:b\mathcal{S} \longrightarrow b\mathcal{B}(S^s)$ (or if you prefer $T_{s,t}:b\mathcal{B}(S^t) \longrightarrow b\mathcal{B}(S^s)$). Finally it is easy to see that if Z is an IBSMP then Z is a right process in the sense of Dynkin (1989b) (see e.g. the proof of Theorem 8.3 in Dynkin (1989a))

Now let E be a Polish space equipped with its Borel σ-field $\mathcal{E}$. If $y\in D(E)$ and $t\in[0,\infty)$, let $W_t(y) = y^t\in D(E)$, and $\mathcal{D}[s,t] = \sigma(W_u:s\leq u\leq t)$. We set $D^t = \{y^t:y\in D=D(E)\}$, $\hat{E} = \{(s,y)\in[0,\infty)\times D:y^s=y\}$ and $\hat{\mathcal{E}}=\mathcal{B}(\hat{E})$. Note that $D=D(E)$, D^t and $\hat{E}$ are all Polish spaces, and that $W.\in D(D(E))$.

Assume $\{P_{s,y}:(s,y)\in\hat{E}\}$ are probabilities on $(D,\mathcal{D})$ satisfying the following hypotheses:

(H)

(H_1) $(s,y) \longrightarrow P_{s,y}(A)$ is $\hat{\mathcal{E}}$-measurable $\forall A\in\mathcal{D}$

(H_2) $P_{s,y}(W_s=y) = 1 \quad \forall\ (s,y)\in\hat{E}$

(H_3) If $(s,y)\in\hat{E}$ and $T\geq s$ is a $(\mathcal{D}_{t+})$-stopping time then $\forall\psi\in\mathcal{D}$ $P_{s,y}(\psi|\mathcal{D}_{T+}) = P_{T,W(T)}(\psi)$, $P_{s,y}$-a.s. on $\{T<\infty\}$.

We stress that (H) will be in force throughout this Section.

In particular $W = (D,\mathcal{D},\mathcal{D}[s,t+],W_t,P_{s,y})$ is an IBSMP in $D^t\subset D(E)$. We let $T_{s,t}:b\mathcal{D}\longrightarrow b\mathcal{B}(D^s)$ denote the inhomogeneous semigroup of W. Note that for $s\geq t$ we may also define

$$T_{s,t}\psi(y) - P_{s,y}(\phi(W(t))) - \phi(y^t) \quad \text{(and hence } T_{s,s}\phi=\phi|_{D^s}).$$

<u>Remark 2.1.1</u> (H_3) implies that if $x\in E$ $(\cong D^0)$, $u>0$, $A\in\mathcal{B}(D^u)$ and

$\psi\in\mathcal{D}$, then

$$\int P_{u,y}(\psi)1_A(y)P_{0,x}(W_u\in dy) = \int P_{u,W(u)}(\psi)1_A(W_u)dP_{0,x}$$
$$= P_{0,x}(\psi 1_A(W_u)).$$

Therefore $P_{u,y}$ is a regular conditional probability for $P_{0,x}(\cdot|W_u=y)$ and (H) asserts the existence of a nice collection of these regular conditional probabilities given the "initial" $\{P_{0,x}:x\in E\}$.

We now construct the natural time-homogeneous Markov process associated with W. Define $\hat{W}:[0,\infty)\times D \longrightarrow D(\hat{E})$ by

$$\hat{W}(s,y)(t) = \hat{W}_s(y)(t) = (s+t,W_{s+t}(y)).$$

Let $\hat{\mathcal{D}} = \mathcal{B}(D(\hat{E}))$, $\hat{Y}_t(w) = w(t)$ for $w\in D(\hat{E})$ and let $\hat{\mathcal{D}}_t = \sigma(\hat{Y}_u:u\le t)$ denote the canonical filtration on $D(\hat{E})$. For $(s,y)\in\hat{E}$ define a probability $\hat{P}^{(s,y)}$ on $(D(\hat{E}),\hat{\mathcal{D}})$ by

$$\hat{P}^{(s,y)}(A) = P_{s,y}(\hat{W}_s\in A).$$

If $f\in b\hat{\mathcal{E}}$, $t\ge 0$ and $(s,y)\in\hat{E}$, let

(2.1.1) $$\hat{P}_tf(s,y) = \hat{P}^{(s,y)}f(\hat{Y}_t) = P_{s,y}(f(s+t,W(s+t)))$$

so that

(2.1.2) $$\hat{P}_tf(s,y) = T_{s,s+t}(f(s+t,\cdot))(y)$$

and

(2.1.3) $$T_{s,t}g(y) = \hat{P}_{t-s}\hat{g}(s,y)$$

where $g\in b\mathcal{D}$ and $\hat{g}\in b\hat{\mathcal{E}}$ satisfies $\hat{g}(t,y) = g(y)\ \forall y\in D^t$.

It is clear from (2.1.2) and the Borel measurability of $T_{s,t}f(y)$ that $\{\hat{P}_t:t\ge 0\}$ is a semigroup on $b\hat{\mathcal{E}}$.

<u>Proposition</u> <u>2.1.2</u> $\hat{Y} = (D(\hat{E}),\hat{\mathcal{D}},\hat{\mathcal{D}}_{t+},\hat{Y}_t,\hat{P}^{(s,y)})$ is a Borel right process with càdlàg paths, Polish state space $\hat{E}$ and Borel right semigroup $(\hat{P}_t,t\ge 0)$ satisfying $\hat{P}_t1 = 1$. If W is an inhomogeneous Hunt process, $\hat{Y}$ is also a Hunt process.

The routine proof is given in the appendix. We are ready to

state Fitzsimmon's construction of the general superprocess described in the Introduction.

Theorem 2.1.3 (Fitzsimmons (1988),(1989)) Let $Y = (\Omega,\mathcal{F}^o,\mathcal{F}^o_{t+},Y_t,P^x)$ be a Borel right process with a metrizable Lusin state space $(S,\mathcal{S})$ and Borel right semigroup $(P_t,t\geq 0)$ satisfying $P_t 1=1$. Let $c\in bp\mathcal{S}$, $b\in b\mathcal{S}$ and $n(x,du)$ be a measurable kernel from $(S,\mathcal{S})$ to $((0,\infty),\mathcal{B}(0,\infty))$ such that

$$(2.1.4)\qquad \sup_x\int_0^\infty u\wedge u^2 n(x,du) < \infty.$$

Let $\Phi(x,\lambda) = -b(x)\lambda - c(x)\lambda^2 + \int_0^\infty (1-e^{-\lambda u}-\lambda u)n(x,du)$.

(a) If $f\in bp\mathcal{S}$, there is a unique jointly measurable solution $v_t(x) = V_t f(x)$ of

$$(2.1.5)\qquad v_t(x) = P_t f(x) + \int_0^t P_u(\Phi(\cdot,v_{t-u}(\cdot)))(x)du$$

which is uniformly bounded on compact time intervals. Moreover, $0 \leq V_t f(x) \leq e^{\|b^-\|_\infty t}\|f\|_\infty$ and $\{V_t:t\geq 0\}$ forms an (in general non-linear) semigroup on $bp\mathcal{S}$.

(b) There is a unique (in law) $M_F(S)$-valued Borel right process $(\Omega^o,\mathcal{S}^o,\mathcal{S}^o_{t+},\theta_t,X_t,Q_m)$ on the canonical space of right continuous $M_F(S)$-valued paths such that

$$(2.1.6)\qquad Q_m(\exp\{-<X_t,f>\}) = \exp\{-<m,V_t f>\}\quad \forall\ f\in bp\mathcal{S},\ m\in M_F(S).$$

Moreover

$$(2.1.7)\qquad Q_m(<X_t,f>) = <m,P_t^b f>,\quad f\in b\mathcal{S},$$

where

$$P_t^b f(x) = P^x\left(\exp\left\{-\int_0^t b(Y_s)ds\right\}f(Y_t)\right).$$

(c) If Y is a Hunt process then so is X. If Y is a Hunt process and $n=0$, then X has continuous paths.

(d) If S is Polish and Y has càdlàg paths in S, then X has càdlàg paths in $M_F(S)$.

Definition. We call X the (Y,Φ)-superprocess.

Remarks 1. (a) follows from standard Picard iteration arguments and holds if $(P_t t\geq 0)$ is the semigroup associated with a jointly measurable Markov kernel $\{P_t(x,.):t\geq 0,x\in S\}$ on a measurable state space $(S,\mathcal{S})$. See Dynkin (1988, Theorem 3.1) for a very general result when $\Phi(x,\lambda) = -\lambda^2$.

2. In [F] a stronger version of (2.1.4), in which $u\wedge u^2$ is replaced with $u\vee u^2$, is assumed. In Fitzsimmons (1989) it is shown that the results of [F] remain valid under (2.1.4). Theorem 3.5, Corollary 3.6 and Corollary 4.7 of [F] therefore give (b) and (c).

3. The proof of (d) is given below only because it is not stated explicitly in [F].

Proof of 2.1.3(d). Fix $m\in M_F(S)$ and let $\mathcal{S}_t^m = \mathcal{S}_{t+}^o\vee\{Q_m\text{-null sets}\}$. Embed S as a dense G_δ in a compact metric space $\bar{S}$ [DM,III.17], and let $\{f_n:n\in\mathbb{N}\}$ be dense in $C(\bar{S})$. By [F,(3.5)] we may fix ω outside a Q_m-null set such that $\langle X.,f_n\rangle(\omega)$ is càdlàg for all $n\in\mathbb{N}$. It follows easily from Prohorov's theorem on $\bar{S}$ that $X.(\omega)$ is right continuous in $M_F(S)$ with left limits in $M_F(\bar{S})$. Recall that the bounded uniformly continuous functions on S are a convergence-determining class in $M_F(S)$ and that these functions have unique bounded continuous extensions to $\bar{S}$. To prove that $X.(\omega)$ is Q_m-a.s. càdlàg in $M_F(S)$ it therefore suffices to show $X_{t-}(\bar{S}-S) = 0$ for all $t\geq 0$ Q_m-a.s. By the section theorem we must show

(2.1.8) $\quad X_{T-}(\bar{S}-S) = 0 \quad Q_m\text{-a.s.}$

for each bounded $(\mathcal{S}_t^m)$-predictable time $T\geq 0$.

Suppose $0<T\leq t_0$. Since $D(S)$ is Polish, if $\varepsilon>0$ there is a compact set K_ε in S such that $P^m(Y_t \in K_\varepsilon^c \ \exists\ t \le t_0) \le \varepsilon$. Let $\{T_n\}$ announce T $(T_n < T)$. If $\alpha \ge \|b^-\|_\infty$, then arguing as in [F,(3.4),(3.5)], one obtains randomized stopping times $S_n \le t_0$ for Y such that

$$Q_m(e^{-\alpha T_n} X_{T_n}(K_\varepsilon{}^c)) = P^m\left(\exp\left\{-\alpha S_n - \int_0^{S_n} b(Y_u)du\right\} \mathbf{1}(Y(S_n) \notin K_\varepsilon)\right)$$

$$\le \varepsilon$$

by the choice of K_ε and so

$$Q_m(X_{T_n}(\bar{S}-K_\varepsilon)) \le \varepsilon e^{\alpha t_0}.$$

Let $n \to \infty$ and use Fatou's Lemma and the fact that $\bar{S}-K_\varepsilon$ is open in $\bar{S}$ (K_ε is compact) to conclude

$$Q_m(X_{T-}(\bar{S}-K_\varepsilon)) \le \varepsilon\, e^{\alpha t_0}.$$

(2.1.8) follows. ∎

<u>Notation.</u> If $(S,\mathcal{S})$ is a measurable space and $f \in b p\mathcal{S}$, define $e_f : M_F(S) \to [0,1]$ by $e_f(m) = \exp\{-\langle m,f\rangle\}$.

We assume the following hypotheses throughout the remainder of this sub-section:

(B) $\hat{b} \in b\hat{\mathcal{E}}$, $\hat{c} \in bp\hat{\mathcal{E}}$, $\hat{n}((s,y),du)$ is a (measurable) kernel from $(\hat{E},\hat{\mathcal{E}})$ to $((0,\infty),\mathcal{B}(0,\infty))$ such that $\sup_{(s,y)\in\hat{E}} \int_0^\infty u \wedge u^2\ \hat{n}((s,y),du) < \infty$.

Let

$$\hat{\Phi}((s,y),\lambda) = -\hat{b}(s,y)\lambda - \hat{c}(s,y)\lambda^2 + \int_0^\infty (1 - e^{-\lambda u} - \lambda u)\hat{n}((s,y),du).$$

We say that $\hat{\Phi}$ is Markov if there are $b \in b\mathcal{E}$, $c \in bp\mathcal{E}$ and a kernel $n(x,du)$ from $(E,\mathcal{E})$ to $((0,\infty),\mathcal{B}(0,\infty))$ satisfying (2.1.4) such that $\hat{b}(s,y) = b(y(s))$, $\hat{c}(s,y) = c(y(s))$, $\hat{n}((s,y),du) = n(y(s),du)$ and hence if Φ is defined as in Theorem 2.1.3

(2.1.9) $\quad \hat{\Phi}((s,y),\lambda) = \Phi(y(s),\lambda)$.

We now apply Theorem 2.1.3 to $(\hat{Y},\hat{\Phi})$. If $f\in bp\hat{\mathcal{E}}$, let $\hat{V}_t\hat{f}(s,y)$ denote the unique solution of $(2.1.\hat{5})$ ((2.1.5) with $\hat{P}_t$ and $\hat{\Phi}$ in place of P_t and Φ), and let $\hat{X} = (\hat{\Omega},\hat{\mathcal{G}},\hat{\mathcal{G}}_{t+},\hat{\theta}_t,\hat{X}_t,\hat{Q}_m)$ denote the $(\hat{Y},\hat{\Phi})$-superprocess on the canonical space of càdlàg $M_F(\hat{E})$-valued paths. (By Theorem 2.1.3(d) $\hat{X}$ is càdlàg since $\hat{Y}$ is.) Therefore

(2.1.10) $\hat{Q}_m(e_f(\hat{X}_t)) = \exp\{-\langle m,\hat{V}_tf\rangle\}\ \forall\ f\in bp\hat{\mathcal{E}},\ m\in M_F(\hat{E})$.

On occasion it will be convenient to work with the homogeneous càdlàg Borel right process $\hat{X}$ but the process of principle interest is the IBSMP taking values in the simpler space $M_F(D)$ which is obtained from $\hat{X}$ by a trivial projection.

Notation. $\pi:\hat{E} \longrightarrow D(E)\equiv D$ is the projection $\pi(s,y) = y$. Define $\bar{\pi}:M_F(\hat{E}) \longrightarrow M_F(D)$ by $\bar{\pi}(\nu) = \nu(\pi^{-1}(.))$. (This $^-$ notation will also be used for other mappings.) π and $\bar{\pi}$ are both continuous. Let $\Omega = D(M_F(D))$ with its Borel σ-field, $\mathcal{G}$. $H_t(\omega) = \omega(t)$ $(\omega\in\Omega)$ and let

$$\mathcal{G}[r,s] = \sigma(H_t:r\le t\le s),\quad 0\le r\le s.$$

$\theta_t(\omega)(s) = \omega(t+s)$ are the usual shift operators on Ω. Let

$M_F(D)^s = \{m\in M_F(D):m((D^s)^c) = 0\} \subset M_F(D)$ and

$\hat{M} = \{(s,m)\in[0,\infty)\times M_F(D):m\in M_f(D)^s\}$.

Let $\hat{H}_t = \bar{\pi}(\hat{X}_t)$ so that $\hat{H}.$ is a càdlàg $M_F(D)$-valued process on $\hat{\Omega}$. If $s\ge0$, let $\hat{H}^{(s)}(t) = \hat{H}((t-s)^+)$, $t\ge0$.

Lemma 2.1.4. (a) $\hat{X}_t = \delta_{s+t}\times\hat{H}_t$ and

$$\hat{H}_t\in M_F(D)^{s+t}\ \ \forall\ t\ge0,\ \hat{Q}_{\delta_s\times m}\text{-a.s. } \forall(s,m)\in\hat{M}.$$

(b) If $t\ge0$ is fixed and $f,g\in bp\hat{\mathcal{E}}$ satisfy $f(t,.)=g(t,.)$ on D^t, then $\hat{V}_{t-s}f(s,y) = \hat{V}_{t-s}g(s,y)\ \forall(s,y)\in\hat{E},\ s\le t$.

Proof. (a) Fix $(s,m)\in\hat{M}$ and for $t\ge0$, let $A(t) = \{(u,y)\in\hat{E}:u\ne t+s\}$. Then

$$\hat{Q}_{\delta_s \times m}(\hat{X}_t(A(t)) \le \exp\{\|\hat{b}^-\|t\} \hat{P}^{\delta_s \times m}(\hat{Y}(t) \notin \hat{A}(t)) \quad \text{(by (2.1.7))}$$

$$= 0 \quad \text{(by (2.1.1))}.$$

This gives $\hat{X}_t = \delta_{s+t} \times \hat{H}_t$ $\hat{Q}_{\delta_s \times m}$-a.s. for t fixed and hence for all $t \ge 0$ a.s. by the right continuity of both sides. Since $\hat{X}_t \in M_F(\hat{E})$ $\forall t \ge 0$, it is then obvious that $\hat{H}_t \in M_F(D)^{s+t}$ $\forall t \ge 0$, $\hat{Q}_{\delta_s \times m}$-a.s.

(b) $\exp\{-\hat{V}_{t-s} f(s,y)\} = \hat{Q}_{\delta_s \times \delta_y}(\exp(-\langle\hat{X}_{t-s}, f\rangle))$ (by (2.1.10))

$= \hat{Q}_{\delta_s \times \delta_y}(\exp\{-\langle\hat{H}_{t-s}, f(t,.)\rangle\})$ (by (a)).

If $f(t,.) = g(t,.)$, the right side is unchanged if f is replaced by g and so the same is true for the left side. ∎

<u>Definition</u> If $(s,m) \in \hat{M}$, define a probability $Q_{s,m}$ on $(\Omega, \mathcal{G}[s,\infty))$ by

(2.1.11) $\quad Q_{s,m}(H \circ \theta_s \in A) = \hat{Q}_{\delta_s \times m}(\hat{H} \in A), \quad A \in \mathcal{G}$

or equivalently

(2.1.12) $\quad Q_{s,m}(H \in B) = \hat{Q}_{\delta_s \times m}(\hat{H}^{(s)} \in B), \quad B \in \mathcal{G}[s,\infty)$.

At long last, and after considerable notational gyrations, we arrive at the main result of this section. It should be obvious from the properties of $\hat{X}$ given in Theorem 2.1.3, and Lemma 2.1.4 but we give a proof in the Appendix. Recall our standing assumptions are (H) and (B).

<u>Theorem</u> <u>2.1.5</u> (a) $H = (\Omega, \mathcal{G}, \mathcal{G}[s,t+], H_t, Q_{s,m})$ is an inhomogeneous Borel strong Markov process with càdlàg paths in $M_F(D)^t \subset M_F(D)$.

(b) If W is an inhomogeneous Hunt process, so is H.

(c) If W is an inhomogeneous Hunt process and $\hat{n} = 0$, then H. is $Q_{s,m}$-a.s. continuous on $[s,\infty)$ $\forall$ $(s,m) \in \hat{M}$.

(d) If $f\in b\mathcal{D}$, then $Q_{s,m}(\langle H_t,f\rangle) = \langle m,T^{\hat{b}}_{s,t}f\rangle$, where $T^{\hat{b}}_{s,t}f(y) = P_{s,y}\left(\exp\left\{-\int_s^t \hat{b}(u,W_u)du\right\}f(W_t)\right)$, for $(s,m)\in\hat{M}$, $t\geq s$, $y\in D^s$.

Definition. Let $Q_{s,t}f(m) = Q_{s,m}(f(H_t))$ $(s\leq t$, $(s,m)\in\hat{M}$, $f\in b\mathcal{B}(M_F((D)^t))$ denote the semigroup of H. Hence $Q_{s,t}:b\mathcal{B}(M_F(D)^t) \longrightarrow b\mathcal{B}(M_F(D)^s)$. Let $q(s,m;t,A) = Q_{s,m}(H_t\in A)$ denote the transition function of H.

We now reinterpret (2.1.5) in the time-inhomogeneous setting of H.

Lemma 2.1.6. Let $f\in bp\hat{\mathcal{E}}$. Suppose $v_{s,t}(y)$ is bounded and Borel measurable on $\{(s,y,t)\in\hat{E}\times[0,\infty):t\geq s\}$ and $\hat{v}_{t-s}(s,y) = v_{s,t}(y)$. The following are equivalent:

(2.1.13a) $v_{s,t}(y) = T_{s,t}(f_t)(y) + \int_s^t T_{s,u}(\hat{\Phi}((u,\cdot),v_{u,t}(\cdot)))(y)du$

$\forall y\in D^s$ and $s\leq t$,

(2.1.13b) $\hat{v}_t(s,y) = \hat{P}_t f(s,y) + \int_0^t \hat{P}_u(\hat{\Phi}(\cdot,\hat{v}_{t-u}(\cdot)))(s,y)du$

$\forall (s,y)\in\hat{E}$, $t\geq 0$.

Proof. (2.1.13a) is equivalent to

$$v_{s,t}(y) = \hat{P}_{t-s}f(s,y) + \int_s^t \hat{P}_{u-s}(\hat{\Phi}(\cdot,\hat{v}_{t-u}(\cdot)))(s,y)du$$

$\forall y\in D^s$, $s\leq t$ (by (2.1.3))

$$\Leftrightarrow \hat{v}_{t-s}(s,y) = \hat{P}_{t-s}f(s,y) + \int_0^{t-s} \hat{P}_u(\hat{\Phi}(\cdot,\hat{v}_{t-s-u}(\cdot)))(s,y)du$$

$\forall y\in D^s$, $s\leq t$

$\Leftrightarrow$ (2.1.13b). ∎

Definition. If $f\in bp\mathcal{D}$ define $\hat{f}\in bp\hat{\mathcal{E}}$ by $\hat{f}(s,y) = f(y)$ and for $s\leq t$ and $y\in D^s$, let

(2.1.14) $V_{s,t}f(y) = \hat{V}_{t-s}\hat{f}(s,y)$,

(recall $\hat{V}_t\hat{f}$ is the unique solution of $(2.1.\hat{5})$).

(2.1.15) If $t\geq 0$ is fixed, Lemma 2.1.4(b) shows that (2.1.14) will hold for all $(s,y)\in\hat{E}$ and $s\leq t$ if $\hat{f}\in bp\hat{\mathcal{E}}$ satisfies $\hat{f}(t,y) =$

$f(y)\quad \forall y\in D^t$.

<u>Theorem 2.1.7.</u> (a) If $f\in bp\mathcal{D}$, $V_{s,t}f(y) = v_{s,t}(y)$ satisfies

$$(2.1.16)\quad v_{s,t}(y) = T_{s,t}f(y) + \int_s^t T_{s,u}(\hat{\Phi}((u,.),v_{u,t}(.)))(y)du$$

$y\in D^s$, $s\le t$, is Borel measurable in $(s,y,t)\in\{(s,y,t):(s,y)\in\hat{E},t\ge s\}$, $0 \le V_{s,t}f(y) \le e^{\|\hat{b}^-\|_\infty(t-s)}\|f\|_\infty$ and $V_{s,t}:bp\mathcal{D} \longrightarrow bp\mathcal{B}(D^s)$ forms a nonlinear semigroup. If t is fixed and $v_{s,t}(y)$ is $\hat{\mathcal{E}}$-measurable in $(s,y)\in\hat{E}$, $s\le t$, and satisfies (2.1.16) for $y\in D^s$ and $s\le t$, then $v_{s,t}(y) = V_{s,t}f(y)$ for $y\in D^s$ and $s\le t$.

(b) $\{Q_{s,t}:0\le s\le t<\infty\}$ is the unique Markov semigroup,

$$Q_{s,t}:b\mathcal{B}(M_F(D)^t) \longrightarrow b\mathcal{B}(M_F(D)^s) \text{ such that}$$

$$(2.1.17)\quad Q_{s,t}(e_f)(m) = \exp\{-\langle m,V_{s,t}f\rangle\}\quad \forall f\in bp\mathcal{D},\ m\in M_F(D)^s,\ s\le t.$$

<u>Remark.</u> If $\hat{\Phi}((u,y),\lambda) = -\gamma\lambda^2/2$ for some $\gamma>0$, then, comparing (2.1.16) with (1.6) and (1.7) of Dynkin (1989b), we see that $q(s,m;t,.)$ is precisely the transition function of the superprocess which Dynkin (1989b) associates with the inhomogeneous Markov process W.

<u>Proof.</u> (a) (2.1.$\hat{5}$) and Lemma 2.1.6 show that $V_{s,t}f(y)$ satisfies (2.1.16). Suppose $t\ge 0$ is fixed and $v_{s,t}(y)$ solves (2.1.16) for $s\le t$ and $y\in D^s$. If

$$v_{s,u}(y) = \begin{cases} v_{s,t}(y) & \text{if } u=t \\ 0 & \text{if } u\ne t \end{cases} \quad\text{and}\quad f(s,y) = \begin{cases} f(y) & \text{if } s=t \\ 0 & \text{if } s\ne t \end{cases}$$

then $v_{s,u}$ solves (2.1.13a) and so $v_{s,u}(y) = \hat{V}_{u-s}f(y)$ by (2.1.13b) and uniqueness in (2.1.$\hat{5}$). In particular $v_{s,t}(y) = \hat{V}_{t-s}f(y) = V_{s,t}f(y)$ (the latter by (2.1.15)). The remaining properties of $V_{s,t}f$ are clear from the corresponding properties of $\dot{V}$ (see Theorem 2.1.3(a)).

(b) $Q_{s,t}(e_f)(m) = \hat{Q}_{\delta_s\times m}(\exp\{-\langle\hat{H}_{t-s},f\rangle\})$

$$= \hat{Q}_{\delta_s \times m}(\exp\{-\langle \hat{X}_{t-s}, f\rangle\})$$

$$= \exp\{-\langle m, \hat{V}_{t-s}\hat{f}(s,.)\rangle\}$$

$$= \exp\{-\langle m, V_{s,t}f\rangle\}.$$

The uniqueness of $Q_{s,t}$ follows via the usual monotone class arguments. ∎

<u>Definition.</u> An inhomogeneous Borel strong Markov process $H = (\Omega^o, \mathcal{F}^o, \mathcal{F}^o[s,t+], H_t, \tilde{P}_{s,m})$ with càdlàg paths in $M_F(D)^t \subset M_F(D)$ whose semigroup is given by (2.1.16),(2.1.17) is called the $(W,\hat{\Phi})$-historical process, or the $(P_{s,y},\hat{\Phi})$-historical process. If $\Pi_t:D(E) \longrightarrow E$ is the projection map $\Pi_t(y) = y(t)$ and $\bar{\Pi}_t:M_F(D) \longrightarrow M_F(E)$ is given by $\bar{\Pi}_t(\nu) = \nu(\Pi_t^{-1}(.))$, then $X_t = \bar{\Pi}_t(H_t)$ is the $(W,\hat{\Phi})$- (or $(P_{s,y},\hat{\Phi})$-) superprocess.

Clearly the laws of H_t and X_t (under a given $\tilde{P}_{s,m}$) are unique and so nothing is lost in working with the canonical version of H constructed in Theorem 2.1.5 under (H) and (B), which is our setting for most of this paper.

<u>Theorem 2.1.8.</u> $\forall (s,m)\in\hat{M}$, $Q_{s,m}$-a.s. X_t has càdlàg $M_F(E)$-valued paths and $X_{t-} = \bar{\Pi}_t(H_{t-})$ $\forall t\geq s$.

<u>Proof.</u> Fix $(s,m)\in\hat{M}$ and ω outside a $Q_{s,m}$-null set such that $H_u\in M_F(D)^u$ $\forall u\geq s$. Let $T>t$. Then for H_t-a.a. y, $y(.)$ is continuous at T. Since $(D^t)^c$ is open in D, $H_{t-}((D^t)^c) \leq \underline{\lim}_{s\uparrow t} H_s((D)^t)^c) = 0$ and therefore $y(.)$ is continuous at T for H_{t-}-a.a. y as well. Therefore H_{t-}-a.a. and H_t-a.a. paths y are continuity points for the map Π_T. If $h:E \longrightarrow \mathbb{R}$ is bounded and continuous, then $(s<t)$

$$\langle h, X_s\rangle = \langle h\circ\Pi_s, H_s\rangle = \langle h\circ\Pi_T, H_s\rangle$$

$$\longrightarrow \langle h\circ\Pi_T, H_{t-}\rangle \text{ as } s\uparrow t$$

$$= \langle h\circ\Pi_t, H_{t-}\rangle \quad (\text{since } H_{t-}((D^t)^c) = 0)$$

$$= \langle h, \bar{\Pi}_t(H_{t-})\rangle.$$

Hence $X_{t-} = \bar{\Pi}_t(H_{t-})$. The obvious modification of the above argument then shows the right-continuity of X. ∎

The historical process allows us to canonically decompose H_t into a sum of independent historical processes according to the initial state of their ancestors. The analogous decomposition for the nonstandard model (for Brownian motion) was used in Perkins (1990, Section 4).

<u>Theorem 2.1.9.</u> Let $(s,m) \in \hat{M}$ and let $\{C_j : j \in \mathbb{N}\} \subset \mathcal{D}$ be disjoint sets whose union supports m. Define $H_t^j \in M_F(D)$ by $H_t^j(A) = H_t(A \cap C_j^s)$ where $C_j^s = \{y : y^s \in C_j\}$. Under $Q_{s,m}$, $\{H^j : j \in \mathbb{N}\}$ are independent càdlàg $M_1(D)$-valued inhomogeneous $\{\mathcal{G}[s,t+) : t \ge s\}$-strong Markov processes such that

$$(2.1.18) \quad Q_{s,m}(H^j \in A) = Q_{s,m_j}(H \in A), \quad \sum_{j=1}^{\infty} H_t^j = H_t \ \forall t \ge s, \ Q_{s,m}\text{-a.s.},$$

where $m_j(B) = m(B \cap C_j)$.

<u>Proof.</u> If $\phi \in C_b(D)$ then since $1_{C_j}(w_s)\phi(w_t)$ is càdlàg in $t \in [s,\infty)$, Theorem 3.5 of [F] and the definition of $Q_{s,m}$ imply that $\langle H_t^j, \phi\rangle = \langle H_t, 1_{C_j^s}\phi\rangle$ is càdlàg on $[s,\infty)$ $Q_{s,m}$-a.s. This shows H^j is a càdlàg $M_F(D)$-valued process under $Q_{s,m}$.

If $f_j \in bp\mathcal{D}$, the uniqueness in (2.1.16) implies

$$(2.1.19) \quad V_{u,t}(1_{C_1^s}f_1 + 1_{C_2^s}f_2) = 1_{C_1^s}V_{u,t}f_1 + 1_{C_2^s}V_{u,t}f_2, \ s < u \le t.$$

If $s \le t_1 \le t_2$ then

$$Q_{s,m}(\exp\{-\langle H_{t_2}^1, f_1\rangle - \langle H_{t_2}^2, f_2\rangle\} | \mathcal{G}[s,t_1+])$$

$$= Q_{s,m}(\exp\{-\langle H_{t_2}, 1_{C_1^s}f_1 + 1_{C_2^s}f_2\rangle\} | \mathcal{G}[s,t_1+])$$

(2.1.20)

$$= \exp\{-\langle H_{t_1}, 1_{C_1^s} V_{t_1,t_2} f_1 + 1_{C_2^s} V_{t_1,t_2} f_2\rangle\}$$

(by (2.1.17) and (2.1.19))

$$= \exp\{-\langle H^1_{t_1}, V_{t_1,t_2} f_1\rangle - \langle H^2_{t_1}, V_{t_1,t_2} f_2\rangle\}.$$

Taking $f_2=0$ we see H^1 is a $\mathcal{G}[s,t+]$-Markov process with the same transition probabilities and initial law as $Q_{s,m_1}(H\in\cdot)$. This gives the first equality in (2.1.18). (2.1.20) and an obvious induction show H^1 and H^2 are independent. The obvious extension gives the independence of $\{H^j\}$.

If T is a $\{\mathcal{G}[s,t+):t\geq s\}$-stopping time, then by the strong Markov property for H

$$Q_{s,m}(H^j_{T+\cdot}\in A|\mathcal{G}[s,T+])$$
$$= Q_{T,H_T}(H^j_\cdot\in A)$$
$$= Q_{T,H^j_T}(H^j_\cdot\in A)$$

(the last equality follows from (2.1.19) with $f_2=0$). This gives the strong Markov property of H^j.

It is clear that $H^{(n)}_t = H_t - \sum_{j=1}^n H^j_t \in M_F(D)$ $\forall t\geq s$, $n\in\mathbb{N}$. If $\beta = \|\hat{b}^-\|_\infty$, then $\{e^{-\beta t}\langle H^{(n)}_t, 1\rangle : t\geq s\}$ is a $Q_{s,m}$-supermartingale by Theorem 2.1.5(d) and so by a weak maximal inequality

$$Q_{s,m}(\sup_{s\leq t\leq T}\langle H^{(n)}_t, 1\rangle \geq e^{\beta T}\delta)$$
$$\leq Q_{s,m}(\sup_{s\leq t\leq T} e^{-\beta t}\langle H^{(n)}_t, 1\rangle \geq\delta)$$
$$\leq (2/\delta)m(\cup_{j>n} C_j) \longrightarrow 0 \quad \text{as} \quad n\longrightarrow\infty.$$

The second part of (2.1.18) follows. ∎

Remark 2.1.10. We have not assumed that $Y_t(y) = y(t)$ $(y \in D)$ is Markov under $\{P_{0,x}: x \in E\}$. In the non-Markovian setting it may sometimes be convenient to replace $D(E)$ with $D(x_0;E) = \{y \in D(E): y(0)=x_0\}$ for a fixed $x_0 \in E$. The above construction goes through essentially unchanged. $P_{s,y}$ are now given probabilities on $D(x_0;E)$ and $H.\in D(M_F(D(x_0;E)))$. Recalling Remark 2.1.1, we see that in this case (H) asserts the existence of a nice collection of regular conditional probabilities $P_{0,x_0}(\cdot|W_u=y)$, $(u,y)\in\hat{E}$, $y \in D(x_0;E)$. P_{0,x_0} is the law of a single (in general non-Markovian) càdlàg process and, assuming (H), we may define a $(P_{0,x_0},\hat{\Phi})$-superprocess.

2.2. The Markov Case.

In what follows we will mostly be concerned with the "Markov setting" i.e. $Y(t)$ is Markov under $\{P_{0,x}: x \in E\}$ and $\hat{\Phi}((s,y),\lambda) = \Phi(y(s),\lambda)$ is Markov. This setting will exhibit a wide class of examples for which (H) is satisfied. In addition we will see that the above definition of $(W,\hat{\Phi})$-superprocess extends the notion of (Y,Φ)-superprocess (see Theorem 2.1.3).

The basic assumptions of this section are

(M_1) $Y = (D,\mathcal{D},\mathcal{D}_{t+},\theta_t,Y_t,P^x)$ is an E-valued Borel right process with càdlàg paths, and

(M_2) $\hat{\Phi}((s,y),\lambda) = \Phi(y(s),\lambda)$ is Markov. (Φ as in Theorem 2.1.3 with $S = E$.)

Notation. If $y,w \in D(E)$ and $s \geq 0$, let $Y(y)(s)=y(s)$ and

$$(y/s/w)(u) = \begin{cases} y(u) & \text{if } u<s \\ w(u-s) & \text{if } u \geq s \end{cases} \in D(E).$$

<u>Theorem 2.2.1.</u> Assume that Y satisfies (M_1). Then

(a) $P_{s,y}(A) = P^{y(s)}(w\in D: y/s/w \in A)$, $(s,y)\in\hat{E}$,

satisfies (H). Hence $W = (D,\mathcal{D},\mathcal{D}[s,t+],W_t,P_{s,y})$ is an IBSMP with càdlàg paths in $D^t \subset D$ and semigroup

(2.2.1) $$T_{s,t}f(y) = P^{y(s)}(f(y/s/Y^{t-s})), \quad (s,y)\in\hat{E},\ t\ge s,\quad f\in b\mathcal{D}.$$

(b) If Y is a Hunt process, then W is an inhomogeneous Hunt process.

<u>Lemma 2.2.2.</u> Let $s\ge0$, $y\in D(E)$. If $T\ge s$ is a $(\mathcal{D}_{t+})$-stopping time, so is $U(w) = T(y/s/w) - s$. If $A\in\mathcal{D}_{T+}$, then $B = \{w:y/s/w \in A\}\in\mathcal{D}_{U+}$.

<u>Proof.</u> Let T and A be as above. Note that it suffices to fix $t>0$ and show

(2.2.2) $$C = \{w\in D: y/s/w \in A,\ U(w) < t\}\in\mathcal{D}_t.$$

(Take $A=D$ to see U is a $(\mathcal{D}_{t+})$-stopping time).

$$\begin{aligned} w\in C &\iff y/s/w\in A \text{ and } T(y/s/w) < t+s \\ &\iff (y/s/w)^{t+s}\in A \text{ and } T((y/s/w)^{t+s}) < t+s \\ &\qquad (\text{by [DM,IV.96(c)] since } A\cap\{T<t+s\}\in\mathcal{D}_{t+s}) \\ &\iff (y/s/w^t)\in A \text{ and } T(y/s/w^t) < t+s \\ &\iff w^t\in C. \end{aligned}$$

A further application of [DM, IV.96(c)] gives (2.2.2). ∎

<u>Proof of Theorem 2.2.1.</u> (H_1) is clear from the Borel measurability of $(s,y,w)\longrightarrow(y/s/w)$ [DM,IV.96(d)] and $x\longrightarrow P^x(A)$. (H_2) is trivial. Turning to (H_3), we fix $(s,y)\in\hat{E}$, $\psi\in b\mathcal{D}$, a $(\mathcal{D}_{t+})$-stopping time $T\ge s$ and $A\in\mathcal{D}_{T+}$ with $A\subset\{T<\infty\}$. Let U and B be as in Lemma 2.2.2. If $g\in b\mathcal{D}$ and $h(w,w') = g(w/U(w)/w')$, then ([DM,IV.96(d)]) h is $\mathcal{D}_{U+}\times\mathcal{D}$ -measurable and the SMP for Y gives

(2.2.3) $$\int h(w,w')P^{Y(U)(w)}(dw') = P^x(h(\cdot,\theta_{U(\cdot)}(\cdot))\,|\,\mathcal{D}_{U+})(w)$$

$$= P^x(g|\mathcal{D}_{U+})(w) \quad \text{on } \{U < \infty\},\ P^x\text{-a.s.}$$

If $g(w) = \psi(y/s/w)$, then

$$\int 1_A\psi\, dP_{s,y} = \int 1_A(y/s/w)\psi(y/s/w)P^{y(s)}(dw)$$

$$= \int 1_B(w)g(w)P^{y(s)}(dw)$$

$$= \int 1_B(w)\int h(w,w')P^{Y(U)(w)}(dw')P^{y(s)}(dw) \quad \text{(by (2.2.3))}$$

$$= \int 1_B(w)\int \psi(y/s/(w/T(y/s/w)-s/w'))\, P^{Y(T(y/s/w)-s)(w)}(dw')P^{y(s)}(dw)$$

$$= \int 1_A(y/s/w)\int \psi((y/s/w)^{T(y/s/w)}/T(y/s/w)/w')\, P^{(y/s/w)(T(y/s/w))}(dw')P^{y(s)}(dw)$$

$$= \int 1_A(w)\int \psi(w^{T(w)}/T(w)/w')P^{w(T(w))}(dw')P_{s,y}(dw)$$

$$= \int 1_A(w)P_{T,W(T)}(\psi)P_{s,y}(dw).$$

This proves (H_3). The rest of (a) is obvious.

(b) Assume Y is quasi-left-continuous, fix $(s,y) \in \hat{E}$, and let $\{T_n\}$ be $(\mathcal{D}_{t+})$-stopping times increasing to T such that $P_{s,y}(T < \infty) = 1$ and $T_n \geq s$. Let $U_n(w) = T_n(y/s/w)-s \uparrow T(y/s/w)-s \equiv U(w)$. Note that $P^{y(s)}(U < \infty) = 1$ and so $Y(U_n) \longrightarrow Y(U)$ $P^{y(s)}$-a.s. This implies

$$y/s/(w^{U_n}) \longrightarrow y/s/(w^U) \quad \text{for } P^{y(s)}\text{-a.s. } w$$

$$\Rightarrow (y/s/w)^{T_n(y/s/w)} \longrightarrow (y/s/w)^{T(y/s/w)} \quad \text{for } P^{y(s)}\text{-a.a. } w$$

$$\Rightarrow W(T_n) \longrightarrow W(T) \quad P_{s,y}\text{-a.s.} \ \blacksquare$$

<u>Remark.</u> If Y is a càdlàg Borel right process as in Theorem 2.2.1, we may apply Theorem 2.2.1 and Proposition 2.1.2 to conclude that $\hat{Y}$ is a Borel right process with càdlàg paths in $\hat{E}$ and semigroup

$$\hat{P}_t g(s,y) = P^{y(s)}(g(s+t,y/s/Y^t)), \quad g \in b\hat{\mathcal{E}},\ (s,y) \in \hat{E}.$$

<u>Theorem 2.2.3.</u> Assume Y and W are as in Theorem 2.2.1. and

that $\hat{\Phi}$ satisfies (M_2). Then

(a) The $(W,\hat{\Phi})$-historical process $(\Omega,\mathcal{G},\mathcal{G}[s,t+],H_t,Q_{s,m})$ exists. The semigroup, $Q_{s,t}$, of H is uniquely determined by

(2.2.4) $$Q_{s,t}(e_f)(m) = \exp\{-\langle m,V_{s,t}f\rangle\} \quad \forall\ f \in bp\mathcal{D},\ m \in M_F(D)^s,\ s\le t,$$

where $V_{s,t}f(y) = v_{s,t}(y)$ is the unique solution of

(2.2.5) $$v_{s,t}(y) = P^{y(s)}(f(y/s/Y^{t-s})) + \int_0^{t-s} P^{y(s)}(\Phi(Y(u),v_{u+s,t}(y/s/Y^u)))du\ , \quad s\le t,\ y \in D^s$$

which is Borel measurable in $(s,y,t) \in \{(s,y,t) \in \hat{E}\times[0,\infty):t\ge s\}$ and is bounded if $t-s$ is bounded. Moreover if t is fixed, then $(s,y) \longrightarrow V_{s,t}f(y)$ is the unique Borel measurable and bounded solution of (2.2.5) on $\hat{E}\cap([0,t]\times D)$.

(b) If Y is a Hunt process, then H is an inhomogeneous Hunt process.

(c) If Y is a Hunt process and $n = 0$, then $H_\cdot$ is continuous on $[s,\infty)$ $Q_{s,m}$-a.s. $\forall\ (s,m) \in \hat{M}$.

Proof. We claim (2.2.5) is equivalent to (2.1.16) in the present context. Using (2.2.1) we see that (2.1.16) becomes (for a given value of t)

$$v_{s,t}(y) = P^{y(s)}(f(y/s/Y^{t-s})) + \int_s^t P^{y(s)}(\hat{\Phi}((u,y/s/Y^{u-s}),v_{u,t}(y/s/Y^{u-s}))du$$

$$\Leftrightarrow \quad v_{s,t}(y) = P^{y(s)}(f(y/s/Y^{t-s})) + \int_s^t P^{y(s)}(\Phi(Y(u-s),v_{u,t}(y/s/Y^{u-s}))du$$

which is equivalent to (2.2.5).

(a) is now a consequence of Theorems 2.1.5(a), 2.1.7 and 2.2.1.

(b) follows from Theorems 2.1.5(b) and 2.2.1(b).

(c) follows from Theorems 2.1.5(c) and 2.2.1(b). ∎

<u>Note.</u> If Y and W are as in Theorem 2.2.1 and $\hat{\Phi}((s,y),\lambda) = \Phi(y(s),\lambda)$ is Markov we will also refer to the $(W,\hat{\Phi})$-historical process as the (Y,Φ)-historical process.

<u>Theorem</u> <u>2.2.4.</u> Assume Y and W are as in Theorem 2.2.1, $\hat{\Phi}((s,y),\lambda) = \Phi(y(s),\lambda)$ is Markov and $X_t = \bar{\Pi}_t(H_t)$ is the $(W,\hat{\Phi})$-superprocess, where H is the (Y,Φ)-historical process on its canonical path space. If $(s,m) \in \hat{M}$, $T\geq s$ is a $\{\mathcal{G}[s,t+]:t\geq s\}$-stopping time and $\psi \in b\mathcal{B}(D(M_F(E)))$, then

(2.2.6) $$Q_{s,m}(\psi(X\circ\theta_T)|\mathcal{G}[s,T+]) = Q_{X(T)}(\psi) \quad Q_{s,m}\text{-a.s. on } \{T<\infty\}$$

where Q_ν is the law of the (Y,Φ)-superprocess starting at ν (see Theorem 2.1.3). In particular, if we identify E with D^o, then $\forall m \in M_F(E)$, X under $Q_{0,m}$ is equal in law to the (Y,Φ)-superprocess starting at m.

<u>Proof.</u> Let $g \in bp\mathcal{E}$ and let $U_t g(x)$ denote the unique solution of (2.1.5) where P_t is the semigroup of Y. If $(s,y) \in \hat{E}$ and $t\geq s$, let $v_{s,t}(y) = U_{t-s}g(y(s))$ and $f_t(y) = g(y(t))$. Then from (2.1.5) we have

$$\begin{aligned} v_{s,t}(y) &= P_{t-s}g(y(s)) + \int_0^{t-s} P^{y(s)}(\Phi(Y(u),U_{t-s-u}g(Y(u)))du \\ &= P^{y(s)}(f_t(y/s/Y^{t-s})) \\ &\qquad + \int_0^{t-s} P^{y(s)}(\Phi(Y(u),v_{u+s,t}(y/s/Y^u)))du. \end{aligned}$$

Apply the uniqueness in (2.2.5) for a fixed t to conclude

(2.2.7) $$V_{s,t}f_t(y) = v_{s,t}(y) = U_{t-s}g(y(s)) \ \forall\ (s,y) \in \hat{E},\ s\leq t.$$

Let $(s,m) \in \hat{M}$ and assume $T\geq s$ is as in the statement of the Theorem. Then

$$\begin{aligned} &Q_{s,m}\Big(\exp\{-\langle X_{T+t},g\rangle\}\,|\,\mathcal{G}[s,T+]\Big)(w) \\ &= Q_{T(w),H(T)(w)}(\exp\{-\langle X_{T(w)+t},g\rangle\}) \quad \text{(strong Markov property} \end{aligned}$$

for H)

$= Q_{T(w),H(T)(w)}(\exp\{-\langle H_{T(w)+t}, f_{T(w)+t}\rangle\})$

$= \exp\{-\langle H(T)(w), V_{T(w),T(w)+t} f_{T(w)+t}\rangle\}$ (by (2.2.4))

$= \exp\{-\langle H(T)(w), U_t g\circ\Pi_{T(w)}\rangle\}$ (by (2.2.7)

$= \exp\{-\langle X(T)(w), U_t g\rangle\}$

$= Q_{X(T)(w)}(e_g(x_t))$,

where sample paths in $D(M_F(E))$ are denoted by $x_\cdot$.

It is now easy to obtain (2.2.6) first for

$\psi(x) = \exp\left\{-\sum_{i=1}^{n} \langle x(t_i), g_i\rangle\right\}$ ($g_i \in bp\mathcal{E}$) by using the SMP at $T+t_i$ and then for all $\psi \in \mathcal{B}(D(M_F(E)))$ by a monotone class argument (e.g.[DM,I.21]). The last statement of the Theorem follows by taking T=0. ∎

Note that (2.2.1) and (2.2.7) yield the following relations if $f_t(y) = g(y(t))$, $g\in bp\mathcal{E}$,

(2.2.8) $T_{s,t}f_t(y) = P_{t-s}g(y(s))$

(2.2.9) $V_{s,t}f_t(y) = U_{t-s}g(y(s))$ $\forall(s,y)\in \hat{E}$, $s\le t$.

<u>Notation.</u> $U_t^{(n)}: bp\mathcal{E}^n \longrightarrow bp\mathcal{E}^{n-1}$ is defined by

$(U_t^{(n)}g)(x_1,\ldots,x_{n-1}) = U_t g(x_1,\ldots,x_{n-1},\cdot)(x_{n-1})$.

We define $\bar{U}_t^{(2)}: bp\mathcal{D}[0,s+]\times\mathcal{E} \longrightarrow bp\mathcal{D}[0,s+]$ by

$\bar{U}_t^{(2)}g(y) = U_t g(y^s,\cdot)(y(s))$.

<u>Theorem</u> <u>2.2.5.</u> (a) Let $f_{s,t}(y) = g(y,y(t))$, $g\in bp\mathcal{D}[0,s+]\times\mathcal{E}$. Then

(2.2.10) $V_{s,t}f_{s,t}(y) = (\bar{U}_{t-s}^{(2)}g)(y) \in \mathcal{D}[0,s+]$

$\forall(s,y)\in \hat{E}$, $s\le t$.

(b) If $g\in bp\mathcal{E}^n$, and $t_1\le t_2\le\ldots\le t_n$, and $f_{t_1,\ldots,t_n}(y) = g(y(t_1),\ldots,y(t_n))$, then for any $1\le k\le n$,

(2.2.11) $V_{t_k,t_n} f_{t_1,\ldots,t_n}(y)$

$$= (U^{(k+1)}_{t_{k+1}-t_k} \cdots U^{(n)}_{t_n-t_{n-1}} g)(y(t_1),y(t_2),\ldots,y(t_k)).$$

(c) The process H_t has the following "extended" Markov property: for $\psi \in b\mathcal{G}$

$$(2.2.12) \quad Q_{u,m}(\psi(\bar{\theta}_s(H_{u+.}))) = Q_{u-s,\bar{\theta}_s(m)}(\psi(H_{u-s+.})) \qquad \text{for all } s \le u,\ (s,m) \in \hat{M}.$$

(Here $\bar{\theta}_s(H_{u+.})(t) = \bar{\theta}_s(H_{u+t}).$)

<u>Proof.</u> (a) For $t \ge s$, $V_{s,t} f_{s,t}(y)$ satisfies

$$v_{s,t}(y) = T_{s,t} f_{s,t}(y) + \int_s^t T_{s,u}(\Phi(y(u),v_{u,t}(\cdot)))du.$$

Now repeat the proof of Theorem 2.2.4. In particular we will show that the right side of (2.2.10) satisfies (2.2.5) . Let $\tilde{v}_{s,t}(y) := U^{(2)}_{t-s} g(y)$. Since

$$(U^{(2)}_{t-s} g)(y) = P_{t-s} g(y,\cdot)(y(s)) + \int_0^{t-s} P_u(\Phi(\cdot,U^{(2)}_{t-s-u}(g(y,\cdot))))(y(s))du$$

it follows from (2.2.8) that

$$\tilde{v}_{s,t}(y) = \\ = P^{y(s)}(f_{s,t}(y/s/Y^{t-s})) + \int_0^{t-s} P^{y(s)}(\Phi(Y(u),\tilde{v}_{u+s,t}(y/s/Y^u)))du.$$

The result follows by uniqueness.

(b) The proof follows by induction using $V_{t_k,t_n} = V_{t_k,t_{n-1}} V_{t_{n-1},t_n}$ and (a).

(c) A simple induction using the Markov property of H shows it suffices to consider $\psi(H) = \psi_0(H_t)$ for some $t \ge 0$ and bounded measurable ψ_0 on $M_F(D)$. If $\psi_0(\mu) = e_{f_{t_1,\ldots,t_n}}(\mu)$ where $0 \le t_1 < \ldots < t_n = u+t-s$ and $t_k = u-s$ then apply (b) and (2.2.4) to see that either side of (2.2.12) equals

$$\exp\left\{- \langle m, U^{(k+1)}_{t_{k+1}-t_k} \cdots U^{(n)}_{t_n-t_{n-1}} g(y(s+t_1),\ldots,y(s+t_k))\rangle\right\}.$$

The result for general ψ_0 then follows by a monotone class argument. ∎

<u>Remark</u> In Dynkin (1988,1989) and Fitzsimmons (1988) more general Markov processes, Y, have been considered. In [F] Y was allowed to be a Borel right process (not necessarily with left limits) whose state space E is a metrizable Lusin space. It is not hard to extend the construction in [F] to the setting where E is a cosouslin metrizable space (i.e. homeomorphic to the complement of an analytic set of a compact metric space) [Fitzsimmons, private communication]. The set $D_+(E)$ of right continuous mappings from $[0,\infty)$ to a cosouslin metrizable space E is itself a cosouslin metrizable space with respect to the topology of pointwise convergence on the rationals [DM,IV.19]. It is then not hard to check that $\hat{E}_+ = \{(s,y)\in[0,\infty)\times D_+(E):y=y^s\}$ is also cosouslin. By applying Fitzsimmons' construction to the obvious analogue of $\hat{Y}$ in this cosouslin state space one may define a Borel right process $\hat{X}_t$ and project down to a historical process H_t if one is given a collection of probabilities $\{P_{s,y}:(s,y) \in \hat{E}_t\}$ on $D_+(E)$ satisfying (H) on the space of right continuous paths. Neither $\hat{X}$ nor H will have left limits, in general, but their other properties remain unchanged. In Theorem 2.2.4 one can then project down the historical process in the Markov setting to arrive at the most general (Y,Φ)-superprocess in [F] (extended to include cosouslin state spaces). The cost of this greater generality is the introduction of a relatively unwieldly topology on $D_+(E)$ and hence on $M_F(D_+(E))$ for H_t. The topology on $M_F(D(E))$ when D(E) is given the J_1-topology is well-studied and has proven to be a

fruitful one in a variety of contexts. As a result we have sacrificed generality for the sake of the down-to-earth choices of Y which will interest us in subsequent sections (such as Lévy processes in $\mathbb{R}^d$).

3. The Probabilistic Structure of H_t.

Having established a general framework for historical processes we now turn to a more detailed study of the IBSMP $H = (\Omega, \mathcal{G}, \mathcal{G}[s,t+], H_t, Q_{s,m})$ in the Markov case. Our objective in this section is to describe H_t in terms of an embedded family of branching particle systems and is closely related to the recent work of Le Jan (1989).

Throughout this section we assume the hypotheses (M_1) and (M_2). Recall that in this case Y is an E-valued Borel right process with càdlàg paths and semigroup P_t, W is the associated IBSMP with càdlàg paths in D^t and inhomogeneous semigroup $T_{s,t}$ and formal generator A_s, H is the $(W,\hat{\Phi})$-historical process with cumulant semigroup $V_{s,t}$, and $X_t = \bar{\Pi}_t(H_t)$ is the (Y,Φ)-superprocess with càdlàg paths in $M_F(E)$ and cumulant semigroup U_t.

As was pointed out in the introduction, for each t H_t is an infinitely divisible random measure. This fact will play an important role in both this section and the next. Central to this development is the canonical representation which we now develop. In order to prepare for the spatially homogeneous case to be considered in Section 6 we include the case of infinite measures. Systematic developments of the theory of random measures can be found in Kallenberg (1977b), Matthes, Kerstan and Mecke (1978) and Liemant, Matthes and Wakolbinger (1988).

As before, (E,d) is a Polish space. Let $\mathcal{E}_b$ the ring of all bounded sets in $\mathcal{E}$. Let $M(E)$ denote the set of all measures on $\mathcal{E}$ and $M_{LF}(E)$ denote the set of all locally finite measures, i.e. measures on $\mathcal{E}$ taking finite values on $\mathcal{E}_b$. Let $C_{bb}(E)$ be the set of all bounded d-continuous mappings with bounded support. Let $\mathcal{M}$ be the σ-field on M_F (or $M_{LF}(E)$) generated by the mappings $\mu \longrightarrow \mu(A)$, $A\in\mathcal{E}_b$. Let τ_w (τ_v) denote the weak topology on $M_F(E)$ (resp. vague on $M_{LF}(E)$), i.e. the topology generated by the mappings $\mu \longrightarrow \langle\mu,\phi\rangle \equiv \int\phi d\mu$, $\phi\in C_b(E)$ (resp. $C_{bb}(E)$). $(M_F(E),\tau_w)$ and $(M_{LF}(E),\tau_v)$ are Polish spaces and $\mathcal{M}$ coincides with the σ-algebra of Borel subsets.

Remark. The standard treatments of random measures (e.g. Kallenberg (1977b)) assume that the space E is locally compact. The theory has been extended to Polish spaces in Matthes, Kerstan and Mecke (1978) and in Liemant, Matthes and Wakolbinger (1988). Most of the basic results carry over to the more general case. The main difference is that there is an additional requirement in verifying tightness. A subset $A \subset M_{LF}(E)$ is relatively compact in the vague topology if and only if for all bounded and closed sets $B \subset E$,

(3.1a) $\sup_{\mu\in A} \mu(B) < \infty$, and

(3.1b) for any $\varepsilon > 0$, there exists a compact subset $K_\varepsilon \subset B$ such that $\sup_{\mu\in A} \mu(B\cap K_\varepsilon^C) < \varepsilon$.

Let $M_{2,F}(E)$ denote the space of all finite measures on $M_F(E)$ and $M_{2,LF}(E)$ denote the set of all measures, ν, on $(M_{LF}(E),\mathcal{M})$ satisfying

$$\nu(\{0\}) = 0, \quad \int (1-e^{-\mu(A)})\nu(d\mu) < \infty \quad \forall \quad A \in \mathcal{E}_b.$$

<u>Theorem 3.1.</u> There is a one-to-one correspondence between infinitely divisible probability distributions, P, on $(M_{LF}(E),\mathcal{M})$ and $(m,R) \in M_{LF}(E) \times M_{2,LF}(E)$ determined by the Laplace functional

$$(3.2) \quad P(e^{-<\mu,\phi>}) = \exp\left(-<m,\phi> - \int (1-\exp(-<\nu,\phi>))R(d\nu)\right), \quad \phi \in p\mathcal{E}_b.$$

The measure R is known as the canonical measure of P, and (3.2) is known as the canonical representation.

<u>Proof.</u> The fact that every infinitely divisible random measure has the representation (3.2) is proved in Kallenberg (1977b, 6.3) when E is locally compact and can be extended without difficulty to the case of Polish spaces (see Liemant et al (1988, Chapt. 1)). The fact that there exists an infinitely divisible random measure with Laplace functional (3.2) when E is Polish (or even Lusin) follows from (A.6) in Fitzsimmons (1988).■

We next describe some basic families of infinitely divisible random measures which are described in terms of their canonical representation.

Given $\Lambda \in M_{LF}(E)$ the Poisson random measure X_Λ with intensity Λ is a $\mathbb{N}$-valued random measure on $(E,\mathcal{E})$ with law $Pois_\Lambda$ uniquely determined by its Laplace functional

$$(3.3) \quad L_\Lambda(\phi) = Pois_\Lambda\left(e^{-<\phi,X_\Lambda>}\right) = e^{-\int (1 - e^{-\phi(x)})\Lambda(dx)}, \quad \phi \in bp\mathcal{E}_b.$$

<u>Remark.</u> We will sometimes refer to an $\mathbb{N}$-valued random measure as a point process or a particle system.

Now let E_1 and E_2 be two Polish spaces. Assume that for $x \in E_1$, $P_x \in M_1(M_{LF}(E_2))$, $x \to P_x$ is measurable, $\Lambda \in M_{LF}(E_1)$ and

$\int P_x\Lambda(dx) \in M_{2,LF}(E_2)$. (Unless otherwise noted we assume that $P_x(\{0\}) = 0$.) If to each point $x\in E_1$ we assign (independently) a random cluster on E_2 with law P_x we have the Poisson cluster measure with Laplace functional

$$(3.4)\qquad L_{\Lambda,\{P_x\}}(\phi) = e^{-\int(1-P_x e^{-\langle\cdot,\phi\rangle})\Lambda(dx)}, \quad \phi\in bp\mathcal{E}_{2,b}.$$

Finally we consider the case in which the intensity Λ is itself a random measure on E_1, i.e. given by a probability measure P_I on $M_{LF}(E_1)$. Then the resulting random measure on E_2 has Laplace functional

$$(3.5)\qquad L_{I,\{P_x\}}(\phi) = \int\exp\Big(-\int(1-P_x e^{-\langle\cdot,\phi\rangle})\Lambda(dx)\Big)P_I(d\Lambda), \quad \phi\in bp\mathcal{E}_{2,b}.$$

and is called the Cox cluster random measure on E_2 with random intensity I on E_1 and clustering mechanism $\{P_x : x\in E_1\}$.

The following conditioning result for a class of Cox random measures plays an important role in this section. Let $x \longrightarrow P_x$ be a measurable mapping from E_1 to $M_1(M_F(E_2))$. Let I be a random measure on E_1 and X be a random measure on $E_1\times E_2$. Assume that the pair (I,X) has joint probability law $P\in M_1(M_F(E_1)\times M_F(E_1\times E_2))$ such that conditioned on I, X is a Poisson cluster random measure with intensity I on E_1 and cluster distribution $\delta_x\times P_x$ on $E_1\times E_2$. (In the case in which E_2 is a one point set X can be identified with a Poisson cluster random measure on E_1 with intensity I and cluster distribution of the form $P_x(A) = \int 1_A(a\delta_x)\mu(da)$ where $\mu \in M_1(0,\infty)$.) Let

$$\mathcal{G}_0 = \sigma\{1_{(0,\infty)}(X(A\times E_2)) : A\in\mathcal{E}_1\}, \quad \mathcal{G}_1 = \sigma\{I(A) : A\in\mathcal{E}_1\}.$$

The set $\{I : I \text{ is nonatomic}\} \in \mathcal{G}_1$ (cf. Cutler (1984, Theorem 2.2.4)). As usual we choose a version of the conditional expectation $E(X|\mathcal{G}_0 \vee \mathcal{G}_1)$ which is almost surely a random measure on E.

An ε-net for E_1 is a countable partition of E_1 consisting of sets in $\mathcal{E}_1$ each having diameter at most ε. The separability of E_1 gives the existence of an ε-net for every $\varepsilon>0$.

<u>Lemma</u> <u>3.2.</u> Let $\mathcal{P}_n = \{A_i^n : i\in\mathbb{N}\}$ be a $1/n$-net for E_1.

(a) P-a.s. on $\{I : I \text{ is nonatomic}\}$

$$E(X|\mathcal{G}_0\vee\mathcal{G}_1) = \int \delta_x \times \left(\int \mu P_x(d\mu)\right)\tilde{X}(dx), \quad \text{and}$$

$$E\left(e^{-\langle X,\phi\rangle}|\mathcal{G}_0\vee\mathcal{G}_1\right) = \exp\left(\int \log\left(\int e^{-\int\phi(x,y)\mu(dy)} P_x(d\mu)\right)\tilde{X}(dx)\right),$$

$$\phi\in bp\mathcal{E}_1\times\mathcal{E}_2.$$

where $\tilde{X}$ is a random measure on E_1 satisfying

$$\tilde{X}(G) = \lim_{n\to\infty}\sum_{i=1}^{\infty} 1(A_i^n\subset G)1_{(0,\infty)}(X(A_i^n\times E_2)) \quad \text{for } G \text{ open.}$$

(b) If I is a.s. nonatomic, then conditioned on I, $\tilde{X}$ is distributed as a Poisson random measure on E_1 with intensity I.

<u>Proof.</u> See Appendix, section 3.

Given a point process N the probability generating functional (PGF) is defined by

$$G_N(f) = E\, e^{\langle N,\log f(x)\rangle}, \qquad f\in bp\mathcal{E}, \quad f > 0,$$

and this is related to the Laplace functional by

$$L_N(\phi) = G_N(e^{-\phi}).$$

We now consider the historical process H_t from the viewpoint of infinitely divisible random measures.

<u>Proposition</u> <u>3.3</u> (a) There exists a Borel measurable mapping $(s,y,t) \longrightarrow R_{s,t}(y,\cdot)$ from $\{(s,y,t)\in \hat{E}\times[0,\infty): t\geq s\}$ to $M_F(M_F(D))$ such that $R_{s,t}(y,\{0\}) = 0$, and

$$(3.6)\quad -\log Q_{s,\delta_y}(\exp(-\langle H_t, f\rangle))$$

$$= V_{s,t}f(y) = \int (1-e^{-\langle \nu, f\rangle}) R_{s,t}(y, d\nu)$$

(b) For $t > s \geq s'$, $m' \in M_F(D)^{s'}$, H_t (under $Q_{s',m'}$) is a Cox cluster random measure on D^t with intensity $H_s(dy) \cdot R_{s,t}(y, M_F(D))$ on D^s where

$$R_{s,t}(y, M_F(D)) = \lim_{\theta \to \infty} (V_{s,t}(\theta 1))(y).$$

A non-zero random cluster of H_t starting at y^s at time s is given by the random measure on D^t with law

$$(3.7)\quad P^*_{s,t;y}\, e^{-\langle \mu, \phi\rangle} \equiv \int e^{-\langle \nu, \phi\rangle} \frac{R_{s,t}(y, d\nu)}{R_{s,t}(s, M_F(D))}.$$

$$= 1 - \frac{V_{s,t}\phi(y)}{R_{s,t}(y, M_F(D))} = \lim_{\eta \downarrow 0} Q_{s,\eta\delta_{y^s}}\left(e^{-\langle H_t, \phi\rangle} \,\middle|\, \langle H_t, 1\rangle > 0\right)$$

(c) $\quad P^*_{s,t;y}(\{\mu : \mu(\{x \in D^t : x^s = y^s\}^c) = 0\}) = 1.$

<u>Proof.</u> From (2.1.17)

$$(3.8)\quad -\log Q_{s,m}(\exp(-\langle H_t, \phi\rangle)) = \int V_{s,t}\phi(y)\, m(dy).$$

Thus H_t is an infinitely divisible random measure on Ω under $Q_{s,m}$ for $m \in M_F(D)^s$. If $m = \delta_y$, $y \in D^s$, $s \leq t$, Theorem 3.1 implies that there is a unique canonical measure $R_{s,t}(y, \cdot) \in M_{2,LF}(D)$ such that

$$(3.9)\quad -\log Q_{s,\delta_y}(\exp(-\langle H_t, \phi\rangle)) = V_{s,t}\phi(y)$$

$$= \int (1-e^{-\langle \nu, \phi\rangle}) R_{s,t}(y, d\nu) \qquad (s,y) \in \hat{E},\ t \geq s.$$

The Borel measurability of $V_{s,t}\phi(y)$ (Theorem 2.2.3) and (3.9) imply the Borel measurability of $(s,y,t) \longrightarrow R_{s,t}(y,\cdot)$. By Lemma 3.4(a,b) (below) $R_{s,t}(y, M_F(D)) < \infty$, $\int \langle \mu, 1\rangle R_{s,t}(y, d\mu) < \infty$ and therefore $R_{s,t}(y,\cdot) \in M_F(M_F(D))$.

(b) From (3.8), (3.9) it follows that the canonical measures associated to $Q_{s,m}$ are given by $\int R_{s,t}(y,d\nu)m(dy)$. Comparing this canonical representation with (3.4) we conclude that H_t with law $Q_{s,m}$ is a Poisson cluster measure with cluster law $\{P^*_{s,t;y}: y\in D^s\}$ characterized by its Laplace functional

$$(3.10)\qquad P^*_{s,t;y}\, e^{-\langle\mu,\phi\rangle} = \int e^{-\langle\mu,\phi\rangle} R_{s,t}(y,d\mu)/R_{s,t}(y,M_F(D))$$

$$= 1 - \frac{V_{s,t}\phi(y)}{R_{s,t}(y,M_F(D))}$$

where (3.9) shows that

$$(3.11)\qquad R_{s,t}(y,M_F(D)) = \lim_{\theta\to\infty} (V_{s,t}\theta 1)(y) < \infty$$

(the last by Lemma 3.4(a)).

The intensity measure of the Poisson cluster process is then given by $R_{s,t}(y,M_F(D))m(dy)$. Then by the Markov property of H_t we obtain

$$(3.12)\qquad -\log Q_{s',m'}(\exp(-\langle H_t,\phi\rangle))$$

$$= -\log Q_{s',m'}\left(\exp\left(-\int (1-P^*_{s,t;y}e^{-\phi})R_{s,t;y}(M_F(D))H_s(dy)\right)\right)$$

which together with (3.5) yields the result except for the last line of (3.7). From (3.9) and (3.10) we obtain

$$\lim_{\eta\downarrow 0} Q_{s,\eta\delta_y}\left(\exp(-\langle H_t,\phi\rangle)\,|\,\langle H_t,1\rangle > 0\right)$$

$$= \lim_{\eta\to 0} \frac{e^{-\eta V_{s,t}\phi(y)} - e^{-\eta R_{s,t}(y,M_F(D))}}{1 - e^{-\eta R_{s,t}(y,M_F(D))}}$$

$$= 1 - \frac{V_{s,t}\phi(y)}{R_{s,t}(y,M_F(D))} = P^*_{s,t;y}(e^{-\langle\mu,\phi\rangle}).$$

(c) Hence $P^*_{s,t;y}(\{\mu:\mu(\{x\in D^t:x^s=y^s\}^c)=0\}) = 1$ since $Q_{s,\eta\delta_y}(\{\mu:\mu(\{x\in D^t: x^s=y^s\}^c)=0\}) = 1$ (e.g. by Theorem 2.1.5(d)) and $\{\mu:\mu(\{x\in D^t:x^s=y^s\}^c)= 0)\}$ is closed in $M_F(D)$. ∎

For the remainder of this section we restrict our attention to the spatially homogeneous Markov case. In particular we assume that

(3.13a) $\Phi(\lambda) = \left[-b\lambda - c\lambda^2 + \int_0^\infty (1-e^{-\lambda u}-\lambda u)n(du)\right]$, $\lambda>0$, $b\in\mathbb{R}$, $c > 0$

with $\int_0^\infty u\wedge u^2 n(du) < \infty$ (so that $\Phi(0+) = 0$).

(3.13b) $\liminf_{\lambda\to\infty} -\Phi(\lambda)/\lambda^{1+\delta} > 0$ for some $\delta > 0$.

<u>Lemma</u> <u>3.4</u>. Let Φ be given by (3.13) with $\Phi \not\equiv 0$. Then

(a) $R_{s,t}(y,M_F(D)) = 1/g(t-s)$

where $0 < g(t) = \lim_{\theta\to\infty} 1/u(t,\theta)$, and $u(t,\theta)$ is the unique non-negative solution of

(3.14) $du(t,\theta)/dt = \Phi(u(t,\theta))$, $u(0,\theta) = \theta$.

For large θ, $u(.,\theta)$ is non-increasing.

(b) The mean measure of H_t is given by

(3.15) $Q_{s,m}(\langle H_t,\phi\rangle) = \iint \langle\nu,\phi\rangle R_{s,t}(y,d\nu)m(dy)$

$= e^{-b(t-s)}\iint P_{s,w}(dy)\phi(y^t)m(dw) = e^{-b(t-s)}\int T_{s,t}\phi(w)m(dw)$

$< \infty$ if $m \in M_F(D)^s$ and $t\geq s$.

The mean cluster size $P^*_{s,t;y}(\langle\mu,1\rangle) = e^{-b(t-s)}g(t-s) \equiv g_b(t-s)$.

(c) Let $\varepsilon > 0$. If $a \equiv -\Phi'(1/g(\varepsilon))$, then $a > 0$ and

(3.16) $\mathcal{G}_\varepsilon(v) = a^{-1}[av - g(\varepsilon)\Phi((1-v)/g(\varepsilon))]$ is a P.G.F.

(Note that in the continuous branching case $\Phi(\lambda) = -\gamma\lambda^2$, $\mathcal{G}_\varepsilon$ corresponds to simple binary branching.)

<u>Proof.</u> (a) By (2.2.9) $(V_{s,t}\theta 1) = U_{t-s}(\theta 1) = u(t-s,\theta)$ where $u(t,\theta)$ satisfies (3.14). Then from Proposition 3.3(a),

$$R_{s,t}(y,M_F(D)) = \lim_{\theta\to\infty} (V_{s,t}\theta 1)(y) = \lim_{\theta\to\infty} u(t-s,\theta).$$

The existence (and uniqueness) of a non-negative solution to (3.14) is established in Watanabe (1968, Proposition 2.2). The fact that it is non-increasing (provided θ is sufficiently large) follows from the fact that $\Phi(\lambda) < 0$ for large λ, the continuity of Φ and uniqueness of solutions to equation (3.14).
The existence of the limit as $\theta \to \infty$ follows by the monotonicity of $u(t,\theta)$ in θ. The fact that $g(t) = \lim_{\theta\to\infty} 1/u(t,\theta) > 0$ follows from (3.13b) by comparison with the equation $du/dt = -u^{1+\delta}$.
(b) (3.15) follows by replacing ϕ by $\theta\phi$ in (3.9), differentiating with respect to θ and evaluating at $\theta=0$ (cf. Fitzsimmons (1988, Proposition 2.7))(see also Theorem 2.1.5(d)). The second statement immediately follows from (3.15), (3.10) and (a).
(c) For $\eta > 0$,

(3.17)
$$\begin{aligned} -\Phi(\eta(1-v)) &= b\eta(1-v) + c\eta^2(1-v)^2 + \int_0^\infty (e^{-\eta u}+\eta u-1)\,n(du) \\ &\quad + v\eta\int_0^\infty u(e^{-\eta u}-1)\,n(du) + \sum_{m=2}^{\infty} \frac{(\eta v)^m}{m!} \int_0^\infty e^{-u\eta}u^m n(du) \end{aligned}$$

It follows from (3.13a) that the coefficients of v^n, $n\ge 2$, are finite and positive and that the series converges for $0\le v\le 1$. The constant term in $-\Phi((1-v)/g(\varepsilon))$ is given by

$$-\Phi(1/g(\varepsilon)) = -\lim_{\theta\to\infty} \Phi(u(\varepsilon,\theta))$$

≥ 0 by the proof of (a) and the continuity of Φ.

Hence it remains to consider the coefficient of v. The linear term in (3.17) (with $\eta = 1/g(\varepsilon)$) is given by $-a/g(\varepsilon) \equiv \Phi'(1/g(\varepsilon))/g(\varepsilon)$ and hence the linear term in (3.16) is 0. Since $\Phi((1-v)/g(\varepsilon))|_{v=1} = 0$, and $\Phi \not\equiv 0$ it follows that $a > 0$ and the proof is complete. ∎

Let $\mathcal{G}_t^{*\varepsilon} = \sigma\{1_{(0,\infty)}(H_t(A)):A\in\mathcal{D}_{t-\varepsilon}\}$ and let $\tilde{H}_t^\varepsilon$ denote a version of the conditional expectation $Q_{s,m}(H_t|\mathcal{G}_t^{*\varepsilon})/g_b(\varepsilon)$ ($m\in M_F(D)^s$) which is a.s. a random measure. Given a measure μ on D^t, let $r_{t-\varepsilon}\mu(A) = \mu(\{w:w^{t-\varepsilon}\in A\})$.

<u>Proposition</u> <u>3.5.</u> Let $0 < \varepsilon < t-s$.

(a) Under $Q_{s,m}$ $r_{t-\varepsilon}H_t$ is a Cox cluster random measure on $D^{t-\varepsilon}$ with intensity $H_{t-\varepsilon}/g(\varepsilon)$ and cluster law $P_y^* = \Xi\delta_y$ where Ξ is a real valued random variable with $P_y^*(e^{-\theta\Xi}) = 1-g(\varepsilon)U_\varepsilon(\theta 1)$.

(b) Assume that

(3.18) $P_{s,y}\times P_{s,y}(W_u^1=W_u^2) = 0 \quad \forall\ u>s,\ y\in D^s$

where W^1,W^2 are i.i.d. copies of W. Then, conditioned on $\mathcal{G}[0,t-\varepsilon]$, $r_{t-\varepsilon}\tilde{H}_t^\varepsilon$ is a Poisson random measure on $D^{t-\varepsilon}$ with intensity $H_{t-\varepsilon}/g(\varepsilon)$. Also conditional on $r_{t-\varepsilon}\tilde{H}_t^\varepsilon$, H_t is the sum of independent non-zero clusters with laws $P^*_{t-\varepsilon,t;y_i}$, one for each atom y_i of $r_{t-\varepsilon}\tilde{H}_t^\varepsilon$, More precisely we have

$$Q_{s,m}(\exp(-\langle H_t,\phi\rangle)|r_{t-\varepsilon}\tilde{H}_t^\varepsilon)$$
$$= \exp\left\{\int \log P^*_{t-\varepsilon,t;y}(e^{-\langle\cdot,\phi\rangle})r_{t-\varepsilon}\tilde{H}^\varepsilon{}_t(dy)\right\} \quad \phi\in p\mathcal{D}.$$

<u>Proof:</u> (a) Let $\phi\in p\mathcal{D}_{t-\varepsilon}$. Then by (3.8) and the Markov property

$$Q_{s,m}(\exp(-\langle H_t,\phi\rangle)|\mathcal{G}[s,t-\varepsilon]) = \exp(-\langle H_{t-\varepsilon},V_{t-\varepsilon,t}\phi\rangle)$$

$$= \exp\left(-g(\varepsilon)^{-1}\int(1-P^*_{t-\varepsilon,t;y}(e^{-<\cdot,1>\phi(y)}))H_{t-\varepsilon}(dy)\right),\ Q_{s,m}\text{-a.s.}$$

(by Proposition 3.3 and Lemma 3.4).

Noting that $P^*_{t-\varepsilon,t;y}e^{-\theta<\cdot,1>} = 1-g(\varepsilon)U_\varepsilon(\theta 1)$ (cf. (3.10)) the result follows from comparison with (3.4).

(b) We apply Lemma 3.2 with $I = H_{t-\varepsilon}/g(\varepsilon)$, $E_1=E_2=D$, and $X(A\times B) = H_t(\{w:(w^{t-\varepsilon},w)\in A\times B\})$. In the terminology of Lemma 3.2 we have $\mathcal{G}_0 = \mathcal{G}^{*\varepsilon}_t$ and $\mathcal{G}_1 = \sigma(H_{t-\varepsilon})$. Proposition 3.3(b),(c) shows that the setting described prior to Lemma 3.2 is in place with $P_y = P^*_{t-\varepsilon,t;y}$. (3.18) and Proposition 4.1.8(b) show that $I = H_{t-\varepsilon}/g(\varepsilon)$ is a.s. nonatomic. Therefore we may use Lemma 3.2 to conclude

$$Q_{s,m}(r_{t-\varepsilon}H_t(A)\,|\,\sigma(H_{t-\varepsilon})\vee\mathcal{G}^{*\varepsilon}_t)/g_b(\varepsilon)$$

$$= \int 1_A(y)\int\mu(D)P^*_{t-\varepsilon,t;y}(d\mu)\tilde{X}(dy)/g_b(\varepsilon)$$

$$= \tilde{X}(A) \quad \text{(Lemma 3.4(b))}.$$

Since $\tilde{X}$ is $\mathcal{G}^{*\varepsilon}_t$-measurable (see the expression for $\tilde{X}$ in Lemma 3.2) this shows $r_{t-\varepsilon}\tilde{H}^\varepsilon_t(A) = \tilde{X}(A)$ a.s. Lemma 3.2(b) and the Markov property of H give the first result. The expression for the conditional Laplace functional of H_t now follows easily from the conditional Laplace functional given in Lemma 3.2 and the equality $r_{t-\varepsilon}\tilde{H}^\varepsilon_t = \tilde{X}$ a.s. ∎

<u>Remark.</u> When the condition (3.18) is not satisfied a result similar to (b) can be formulated but in this case it is necessary to work on an enriched probability space. For example the motion process can be enriched by adding a new spatial dimension so as to satisfy (3.18) and the original process can then be obtained by an appropriate projection. Condition (3.18) will be assumed

throughout the remainder of this section.

Let us now consider the particle system analogue of the historical process. Let W be the IBMSP with inhomogeneous semigroup $T_{s,t}$ as in Section 2. At time $t = 0$, the initial state is given by a finite set of points in $\mathbb{R}^d$. From each of these initial points we construct independent copies of W. Let $\nu \in M_{LF}([0,T))$, $T \le \infty$. Particles alive at time s undergo branching at the times of jumps of independent inhomogeneous Poisson processes with intensity $\nu(ds)$. At the time of branching a particle dies and is replaced by n offspring with probability p_n, $n \ge 0$, each of which starts life at the location of its parent. The offspring generating function is given by

$$\mathcal{G}(z) = \sum_{n=0}^{\infty} p_n z^n , \quad 0 \le z \le 1.$$

We assume that $\sum np_n < \infty$. Between jumps the particles continue the trajectory begun by its parent. Both the motions and deaths of different particles are independent of each other.

Let $N_F(D^t)$ denote the set of finite integer-valued measures on D^t. The historical particle system H_t^* is the càdlàg $N_F(D)$-valued process

$$H_t^* = \sum_i \delta_{y_i^t} \in N_F(D^t),$$

where y_i^t denotes the trajectory of the ith particle alive at time t (stopped at t).

Let $\Omega^* = D(N_F(D))$ with its Borel σ-field $\mathcal{G}^*$, $H_t^*(\omega) = \omega(t)$, and $\mathcal{G}^*[r,s] = \sigma(H_t^*: r \le t \le s)$. If $H_s^* = m \in N_F(D^s)$, then the resulting law of $H_\cdot^*$ on $\mathcal{G}^*[s,\infty)$ is denoted by $Q_{s,m}''$.

If $H_s^* = \delta_y$, $y \in D^s$, then P.G.F. of H_t^* is given by

$$\text{(3.19a)} \qquad G_{s,t}\xi(y) = Q^*_{s,\delta_y}\left(e^{<H_t^*, \log \xi >}\right), \quad \xi \in b\mathcal{D},\ \xi > 0,\ t>s,$$

and moreover if $H_s^* = \sum \delta_{y_i}$, then the PGF of H_t^* is given by

$$\text{(3.19b)} \qquad \prod_i G_{s,t}\xi(y_i).$$

<u>Theorem 3.6.</u> (a) $G_{s,t}\xi$, $0 \le s \le t$, is the unique solution of the equation

$$\text{(3.20)} \qquad G_{s,t}\xi(y) = e^{-\nu((s,t])} T_{s,t}\xi(y) + \int_s^t T_{s,u} e^{-\nu((s,u])}\left[\mathcal{G}(G_{u,t}\xi(y))\right]\nu(du),$$

If $\nu(ds) = \nu(s)ds$, then formally

$$G_{t,t}\xi(y) = \xi(y),$$

$$-\frac{\partial G_{s,t}}{\partial s} = (A_s - \nu(s))G_{s,t} + \nu(s)\mathcal{G}(G_{s,t}).$$

(b) Let $Pois_m$ denote the law of a Poisson random measure on D^s with intensity m. If we assume that H_s^* has distribution $Pois_m$, then H_t^* has Laplace functional

$$\text{(3.21)} \quad \int Q^*_{s,\mu}\left[e^{-<H_t^*,\phi>}\right] Pois_m(d\mu) = \exp\left\{- \int \left(1 - (G_{s,t}e^{-\phi})(y)\right) m(dy)\right\}$$

that is, H_t^* is a Poisson cluster random measure with intensity m and cluster Laplace functional $(G_{s,t}e^{-\phi})(y)$.

(c) The process H_t^* has the extended Markov property

$$Q^*_{u,\mu}(\psi(\bar{\theta}_s(H^*_{u+\cdot}))) = Q^*_{u-s,\bar{\theta}_s(\mu)}(\psi(H^*_{u-s+\cdot}))$$

for all $\psi \in b\mathcal{G}^*$, $s \le u$, $\mu \in N_F(D^u)$.

<u>Proof.</u> (a) For the derivation of (3.20) see Dawson and Ivanoff (1978) and the references therein and Dynkin (1989d). For the uniqueness of the solution to (3.20) see Dynkin (1989d).

(b) This follows from (a) and (3.4).

(c) The proof is similar to that of part (c) of Theorem 2.2.5. ∎

Given $\varepsilon > 0$, let $\{H_t^{*\varepsilon}: t\geq 0\}$ denote the historical branching particle system with branching rate $\nu(ds) = a\, ds$ and offspring P.G.F. $\mathcal{G}_\varepsilon(v)$ where a and $\mathcal{G}_\varepsilon$ are given by (3.16). As above, let $Q^{*\varepsilon}_{s,\delta_y}$ denote the law of $H^{*\varepsilon}_{\cdot}$ started from δ_y at time s. Then $\{H_t^{*\varepsilon}: t\geq 0\}$ is an inhomogeneous $N_F(D)$-valued Markov process with transition function given by

$$(3.22a)\qquad Q^{*\varepsilon}_{s,\sum_i \delta_{y_i}}\left(e^{\langle H_t^{*\varepsilon}, \log \xi\rangle}\right) = \prod_i G_{s,t}\xi(y_i).$$

By Theorem 3.6 $G_{s,t}\xi$ is given by the unique solution of

$$(3.22b)\qquad G_{s,t}\xi(y) = e^{-a(t-s)}T_{s,t}\xi(y)$$

$$+ a\int_s^t T_{s,u}[\mathcal{G}_\varepsilon(G_{u,t}\xi)](y)e^{-a(u-s)}du ,$$

i.e. formally,

$$G_{t,t}(y) = \xi(y), \qquad -\frac{\partial G_{s,t}}{\partial s} = (A_s - a)G_{s,t} + a\mathcal{G}_\varepsilon(G_{s,t}).$$

<u>Proposition 3.7</u> Assume (3.18), $H_s = m\in M_F(D^s)$ and that $H_s^{*\varepsilon}$ is a Poisson random measure with intensity $m/g(\varepsilon)$. Then for $t \geq s+\varepsilon$, both $H^{*\varepsilon}_{t-\varepsilon}$ and $r_{t-\varepsilon}\tilde{H}_t^\varepsilon$ have a Poisson cluster representation of the form (3.4) with $E_1 = D^s$, $E_2 = D^{t-\varepsilon}$, intensity $m/g(\varepsilon)$ and cluster measures

$$P_y(A) = Q^{*\varepsilon}_{s,\delta_y}(H^{*\varepsilon}_{t-\varepsilon}\in A),\ A\in \mathcal{B}(M_F(D^{t-\varepsilon})),\quad y\in D^s$$

(but note that $P_y(\{0\}) > 0$ if $t > s+\varepsilon$).

Proof. Let $\tilde{H}_t^\varepsilon$ be as in Proposition 3.5. Then by Proposition 3.5(b) and (3.5) its Laplace functional is

$$Q_{s,m}(e^{-\langle \tilde{H}_t^\varepsilon,\phi\rangle}) = Q_{s,m}\left(e^{-\langle H_{t-\varepsilon}/g(\varepsilon),(1-e^{-\phi})\rangle}\right)$$

$$= \exp\left(-\int V_{s,t-\varepsilon}((1-e^{-\phi})/g(\varepsilon))(y)m(dy)\right), \qquad \phi \in bp\mathcal{D}_{t-\varepsilon}.$$

From (2.2.5)

$$-\frac{\partial V_{s,t-\varepsilon}}{\partial s} = A_s V_{s,t-\varepsilon} + \Phi(V_{s,t-\varepsilon}), \quad V_{t-\varepsilon,t-\varepsilon} = (1-e^{-\phi})/g(\varepsilon).$$

(Remark: For ease of following the computations we work with the formal differential equations; these can be translated into the corresponding statements in evolution form without difficulty.)

If $v_{s,t-\varepsilon} \equiv 1-g(\varepsilon)V_{s,t-\varepsilon}$, $v_{t-\varepsilon,t-\varepsilon} = e^{-\phi}$, then

$$(3.23) \qquad -\frac{\partial v_{s,t-\varepsilon}}{\partial s} = A_s v_{s,t-\varepsilon} - g(\varepsilon)\Phi((1-v_{s,t-\varepsilon})/g(\varepsilon)).$$

$$= (A_s - a)v_{s,t-\varepsilon} + a\mathcal{G}_\varepsilon(v_{s,t-\varepsilon}).$$

Hence

$$Q_{s,m}(e^{-\langle \tilde{H}_t^\varepsilon,\phi\rangle}) = \exp\left\{-\int (1-v_{s,t-\varepsilon})(y)\,\frac{m(dy)}{g(\varepsilon)}\right\}$$

where $v_{t-\varepsilon,t-\varepsilon} = e^{-\phi}$ and $v_{s,t-\varepsilon}$ satisfies (3.23).

On the other hand by (3.21),

$$\int Q^{*\varepsilon}_{s,\mu}\left(e^{-\langle H^{*\varepsilon}_{t-\varepsilon},\phi\rangle}\right) Pois_{m/g(\varepsilon)}(d\mu)$$

$$= \exp\left(-\int (1-(G_{s,t-\varepsilon}e^{-\phi})(y))\,\frac{m(dy)}{g(\varepsilon)}\right)$$

and $G_{u,t-\varepsilon}e^{-\phi}$ satisfies (3.22b) with $\xi = e^{-\phi}$. Hence $(G_{u,t-\varepsilon}e^{-\phi})(y) = v_{u,t-\varepsilon}(y)$, $s\le u\le t-\varepsilon$, by uniqueness. Recalling (3.22a) we see from the above Laplace transforms that both $r_{t-\varepsilon}\tilde{H}^{\varepsilon}_t$ and $H^{*\varepsilon}_{t-\varepsilon}$ are Poisson cluster random measures with intensity $m/g(\varepsilon)$ and cluster measures $Q^{*\varepsilon}_{s,\delta_y}(H^{*\varepsilon}_{t-\varepsilon}\in\cdot)$. ∎

We now consider the case

(SB) $\Phi(\lambda) = -\gamma\,\lambda^{1+\beta}$, $0<\beta\le 1$.

Then from (3.14) $u(t,\theta)$ satisfies

$$\frac{\partial u(t,\theta)}{\partial t} = -\gamma\, u^{\beta+1}, \quad u(0,\theta) = \theta, \qquad 0<\beta\le 1,$$

and therefore

$$(3.24) \qquad u(t,\theta) = \frac{\theta}{(1+t\beta\gamma\theta^{\beta})^{1/\beta}}, \qquad \lim_{\theta\to\infty} u(t,\theta) = 1/(t\beta\gamma)^{1/\beta},$$

and $g_b(\varepsilon) = g(\varepsilon) = (\varepsilon\beta\gamma)^{1/\beta}$ is the mean total mass of a cluster of age ε.

In addition $a = (\beta+1)/\varepsilon\beta$ and

$$(3.25a) \qquad a\mathcal{G}_{\varepsilon}(s) = \frac{\beta+1}{\varepsilon\beta}\,[s + (1+\beta)^{-1}(1-s)^{1+\beta}].$$

Since $[s+ (1+\beta)^{-1}(1-s)^{1+\beta}] = [s + (1+\beta)^{-1}\sum_{n=0}^{\infty}\binom{1+\beta}{n}(-s)^n]$,

the offspring probabilities are given by

$$(3.25b) \qquad p_n = (1+\beta)^{-1}\binom{1+\beta}{n}(-1)^n, \quad n \ne 1$$

$$p_1 = 0.$$

<u>Remark.</u> In this case $n(du) = \dfrac{\beta(1+\beta)\gamma}{\Gamma(1-\beta)}u^{-(\beta+2)}du$ (cf. Kawazu and Watanabe (1971)).

<u>Lemma 3.8.</u> In the case (SB)

$$Q^{*\varepsilon}_{0,\delta_x}(H^{*\varepsilon}_t > 0) = \left(\frac{\varepsilon}{t+\varepsilon}\right)^{1/\beta}.$$

Proof. Putting $\xi(x) \equiv \theta$, $Q^{*\varepsilon}_{0,\delta_x}(\theta^{\langle H^{*\varepsilon}_t,1\rangle}) = G_{0,t}\theta$ and

$Q^{*\varepsilon}_{0,\delta_x}(\langle H^{*\varepsilon}_t,1\rangle = 0) = \lim_{\theta\to 0} G_{0,t}\theta$. Then by (3.22b) (set s=0 and differentiate with respect to t)

$$\frac{\partial G_{0,t}\theta}{\partial t} = \frac{1}{\varepsilon\beta}(1-G_{0,t}\theta)^{1+\beta}, \quad G_{0,0}\theta = \theta$$

which yields (use (3.24)):

$$G_{0,t}\theta = 1 - \frac{1}{((1-\theta)^{-\beta} + t/\varepsilon)^{1/\beta}}.$$

The result follows by letting $\theta \to 0$. ∎

Theorem 3.9. Assume (3.18) and (SB).

(a) Let $H_0 = m \in M_F(D)^0$. Then $r_0\tilde{H}^\varepsilon_\varepsilon$ is a Poisson random measure with intensity $m/g(\varepsilon)$ and the process $\{r_t\tilde{H}^\varepsilon_{t+\varepsilon} : t \geq 0\}$ is a branching particle system with transition function given by (3.22).

(b) For $0 \leq s < t$, let

$$r_s\tilde{H}^{t-s,x}_t = r_sP^*_{0,t;x}(H_t \mid \mathcal{G}^{*t-s}_t)/g(t-s).$$

(i.e. a non-zero cluster of $r_s\tilde{H}^{t-s}_t$ starting at $x \in \mathbb{R}^d$).

Then the process $s \longrightarrow r_s\tilde{H}^{t-s,x}_t$ $(0 \leq s < t)$ has the same law as a time inhomogeneous branching historical particle system $\{H^{**}_{s,t} : 0 \leq s \leq t\}$ with initial measure $H^{**}_{0,t} = \delta_x$, branching rate at time s equal to $1/(t-s)$ and offspring distribution

$$p_n = \frac{1}{\beta}\binom{1+\beta}{n}(-1)^n, \quad n \geq 2,$$

$$= 0, \text{ otherwise.}$$

Proof.

(a) We will prove that the processes $s \longrightarrow r_{s-\varepsilon}\tilde{H}_s^{\varepsilon}$ (under $Q_{0,m}$) and $s \longrightarrow H^{*\varepsilon}_{s-\varepsilon}$ (under $Q^{*\varepsilon}_{0,Pois_{m/g(\varepsilon)}}$) have the same finite dimensional distributions. By Proposition 3.7 the one-dimensional marginal distributions coincide. Conditioned on $\sigma\{r_{s-\varepsilon}\tilde{H}_s^{\varepsilon}\}$ H_s is the sum of independent non-zero clusters, one for each atom in $r_{s-\varepsilon}\tilde{H}_s^{\varepsilon}$ (cf. Proposition 3.5(b)). That is, if $s_0 \leq s-\varepsilon$, $m' \in M_F(D)^{s_0}$, $\phi \in p\mathcal{D}$

$$Q_{s_0,m'}\left(e^{-\langle H_s,\phi\rangle} \middle| \sigma\{r_{s-\varepsilon}\tilde{H}_s^{\varepsilon}\}\right)$$

$$= \exp\left(\int \log P^{*}_{s-\varepsilon,s;y}(e^{-\langle\cdot,\phi\rangle})\, r_{s-\varepsilon}\tilde{H}_s^{\varepsilon}(dy)\right)$$

where $P^{*}_{s-\varepsilon,s;y}$ is given by (3.10). Therefore, by the Markov property of H_t and the fact that $\sigma\{r_{s-\varepsilon}\tilde{H}_s^{\varepsilon}\} \subset \mathcal{G}[0,s]$, it follows that for $s \leq t$

$$Q_{s_0,m'}\left(e^{-\langle H_t,\phi\rangle} \middle| \sigma\{r_{s-\varepsilon}\tilde{H}_s^{\varepsilon}\}\right)$$

$$= \exp\left(\int \log P^{*}_{s-\varepsilon,s;y}(e^{-\langle\cdot,V_{s,t}\phi\rangle})\, r_{s-\varepsilon}\tilde{H}_s^{\varepsilon}(dy)\right).$$

But by (3.10)

$$P^{*}_{s-\varepsilon,s;y}\, e^{-\langle\cdot,V_{s,t}\phi\rangle} = 1 - \frac{V_{s-\varepsilon,t}\phi(y)}{1/g(\varepsilon)}$$

$$= \left(1 - \left(\frac{\varepsilon}{t-s+\varepsilon}\right)^{1/\beta}\right) + \left(\frac{\varepsilon}{t-s+\varepsilon}\right)^{1/\beta} P^{*}_{s-\varepsilon,t;y}\, e^{-\langle\cdot,\phi\rangle}.$$

Therefore if $\{y_1,\ldots,y_N\} \subset D^{s-\varepsilon}$

(3.26a)

$$Q_{s_0,m}\left(e^{-\langle H_t,\phi\rangle} \mid r_{s-\varepsilon}\tilde{H}^\varepsilon_s = \sum_{i=1}^N \delta_{y_i}\right)$$

$$= \prod_{i=1}^N \left[\left(1 - \left(\frac{\varepsilon}{t-s+\varepsilon}\right)^{1/\beta}\right) + \left(\frac{\varepsilon}{t-s+\varepsilon}\right)^{1/\beta} P^*_{s-\varepsilon,t;y_i}(e^{-\langle\cdot,\phi\rangle})\right]$$

$$= \sum_{k=1}^N \left(1 - \left(\frac{\varepsilon}{t-s+\varepsilon}\right)^{1/\beta}\right)^{N-k} \left(\frac{\varepsilon}{t-s+\varepsilon}\right)^{k/\beta} \sum_{B\in\Gamma_k} \prod_{y_j'\in B} P^*_{s-\varepsilon,t;y_j'}(e^{-\langle\cdot,\phi\rangle})$$

where Γ_k denotes the set of subsets of $\{y_1,\ldots,y_N\}$ containing k elements.

From (3.26a) it follows that if $\{y_1',\ldots,y_k'\} \subset \{y_1,\ldots,y_N\}$, then

$$Q_{s_0,m}\left(e^{-\langle H_t,\phi\rangle} \mid r_{s-\varepsilon}\tilde{H}^\varepsilon_s = \sum_{i=1}^N \delta_{y_i},\ r_{s-\varepsilon}\tilde{H}^{t-s+\varepsilon}_t = \sum_{j=1}^k \delta_{y_j'}\right)$$

(3.26b) $$= \prod_{j=1}^k P^*_{s-\varepsilon,t;y_j'}(e^{-\langle\cdot,\phi\rangle})$$

(3.26c) $$= Q_{s_0,m}\left(e^{-\langle H_t,\phi\rangle} \mid r_{s-\varepsilon}\tilde{H}^{t-s+\varepsilon}_t = \sum_{j=1}^k \delta_{y_j'}\right) .$$

Thus conditioned on $r_{s-\varepsilon}\tilde{H}^\varepsilon_s = \sum_{i=1}^N \delta_{y_i}$ and $r_{s-\varepsilon}\tilde{H}^{t-s+\varepsilon}_t = \sum_{j=1}^k \delta_{y_j'}$

$$r_{t-\varepsilon}\tilde{H}^\varepsilon_t = \sum_{j=1}^k r_{t-\varepsilon}P^*_{s-\varepsilon,t;y_j'}(H_t \mid \mathcal{G}^{*\varepsilon}_t)/g(\varepsilon)$$

and the summands are independent. In Proposition 3.7 the (possibly zero) cluster in $r_{t-\varepsilon}\tilde{H}^\varepsilon_t$ associated with δ_{y_j} in $r_{s-\varepsilon}\tilde{H}^\varepsilon_s$ has law $Q^{*\varepsilon}_{s-\varepsilon,\delta_{y_j}}(H^{*\varepsilon}_{t-\varepsilon}\in\cdot)$ and by (3.26a) the probability that it is non-zero is $(\varepsilon/(t-s+\varepsilon))^{1/\beta}$. Also the probability that a particle

δ_{y_j} in $H^{*\varepsilon}_{s-\varepsilon}$ produces a non-zero cluster in $H^{*\varepsilon}_{t-\varepsilon}$ is $(\varepsilon/(t-s+\varepsilon))^{1/\beta}$ (cf. Lemma 3.8). Therefore the non-zero cluster in $r_{t-\varepsilon}\tilde{H}^{\varepsilon}_t$ associated with $\delta_{y_j} \in r_{s-\varepsilon}\tilde{H}^{t-s+\varepsilon}_t$, namely,

$r_{t-\varepsilon}P^{*}_{s-\varepsilon,t;y_j}(H_t|\mathcal{G}^{*\varepsilon}_t)/g(\varepsilon)$, has law $Q^{*\varepsilon}_{s-\varepsilon,\delta_{y_j}}(H^{*\varepsilon}_{t-\varepsilon}\in\cdot|\langle H^{*\varepsilon}_{t-\varepsilon},1\rangle \neq 0)$.

This completes the proof that the conditional distribution of

$r_{t-\varepsilon}\tilde{H}^{\varepsilon}_t$ given $r_{s-\varepsilon}\tilde{H}^{\varepsilon}_s = \sum_{i=1}^{N}\delta_{y_i}$ and $H^{*\varepsilon}_{t-\varepsilon}$ given $H^{*\varepsilon}_{s-\varepsilon} = \sum_{i=1}^{N}\delta_{y_i}$, (with $s<t$) coincide. We next verify the Markov property of the process $t \longrightarrow r_{t-\varepsilon}\tilde{H}^{\varepsilon}{}_t$. Let $s<t<u$. We will verify that $\{H_u:u\geq t\}$ and consequently $\{r_{u-\varepsilon}\tilde{H}^{\varepsilon}_u:u\geq t\}$ is conditionally independent of $r_{s-\varepsilon}\tilde{H}^{\varepsilon}_s$ given $r_{t-\varepsilon}\tilde{H}^{\varepsilon}_t$. The proof of the general Markov property will then be clear. To verify the above it suffices (because of the Markov property of $H_\cdot$) to prove that H_t is conditionally independent of $r_{s-\varepsilon}\tilde{H}^{\varepsilon}_s$ given $r_{t-\varepsilon}\tilde{H}^{\varepsilon}_t$. (3.26c) implies that if $r = r_{s-\varepsilon}\tilde{H}^{t-s+\varepsilon}_t$ and $r' = r_{s-\varepsilon}\tilde{H}_s - r$, then H_t is conditionally independent of r' given r. The inclusions $\sigma(r) \subset \sigma(r_{t-\varepsilon}\tilde{H}^{\varepsilon}_t) \subset \sigma(H_t)$ allow one to extend the above to conditional independence of H_t and r' given $r_{t-\varepsilon}\tilde{H}^{\varepsilon}_t$ (this general fact is a routine exercise). Now note that

$$Q_{s_0,m}(H_t\in A|r_{s-\varepsilon}\tilde{H}^{\varepsilon}_s, r_{t-\varepsilon}\tilde{H}^{\varepsilon}_t)$$

$$= Q_{s_0,m}(Q_{s_0,m}(H_t\in A|r', r_{t-\varepsilon}\tilde{H}^{\varepsilon}_t)|r+r', r_{t-\varepsilon}\tilde{H}^{\varepsilon}_t)$$

$$\text{(since } \sigma(r) \subset \sigma(r_{t-\varepsilon}\tilde{H}^{\varepsilon}_t))$$

$$= Q_{s_0,m}(H_t\in A|r_{t-\varepsilon}\tilde{H}^{\varepsilon}_t)$$

by the conditional independence established above. Together with the Markov property of $H^{*\varepsilon}_t$ (Theorem [illegible]) this completes the proof that the finite dimensional distributions coincide.

(b) We first show that for the given $0 < \varepsilon < t$, the laws of $r_{t-\varepsilon}\tilde{H}_t^{\varepsilon,x}$ and $H_{t-\varepsilon,t}^{**}$ coincide. By (a) it suffices to show that the laws of $H_{t-\varepsilon}^{*\varepsilon}$ (under $Q_{0,\delta_x}^{*\varepsilon}$), conditioned to be non-zero, and $H_{t-\varepsilon,t}^{**}$ coincide. By Theorem 3.6 the process $\{H_{s,t}^{**}: 0 \le s < t\}$ is a time inhomogeneous Markov process whose transition Laplace functional satisfies

$$E\left(e^{-\langle H_{t-\varepsilon,t}^{**},\phi\rangle} \,\middle|\, H_{s,t}^{**} = \delta_y\right) = \tilde{v}_{s,t-\varepsilon}(y), \text{ and for } 0 \le s \le t-\varepsilon,$$

where $\tilde{v}_{t-\varepsilon,t-\varepsilon} = e^{-\phi}$ and for $0 \le s \le t-\varepsilon$, $\tilde{v}$ is the unique solution of

$$(3.27)\quad -\frac{\partial \tilde{v}_{s,t-\varepsilon}}{\partial s}$$

$$= \left(A_s - \frac{1}{t-s}\right)\tilde{v}_{s,t-\varepsilon} + \frac{1}{(t-s)\beta}\left[(1-\tilde{v}_{s,t-\varepsilon})^{1+\beta} + (1+\beta)\tilde{v}_{s,t-\varepsilon} - 1\right].$$

By (3.22), if $y \in D^s$, $0 < s < t-\varepsilon$,

$$Q_{s,\delta_y}^{*\varepsilon}\left(e^{-\langle H_{t-\varepsilon}^{*\varepsilon},\phi\rangle}\right) = v_{s,t-\varepsilon}(y),$$

where $v_{t-\varepsilon,t-\varepsilon} = e^{-\phi}$, and

$$-\frac{\partial v_{s,t-\varepsilon}}{\partial s}$$

$$= (A_s - (\beta+1)/\varepsilon\beta)v_{s,t-\varepsilon} + \frac{\beta+1}{\varepsilon\beta}[v_{s,t-\varepsilon} + (1+\beta)^{-1}(1-v_{s,t-\varepsilon})^{1+\beta}]$$

We will now find the Laplace functional of $H_{t-\varepsilon}^{*\varepsilon}$ conditioned to be non-zero. By Lemma 3.8,

$$Q_{s,\delta_y}^{*\varepsilon}(\langle H_{t-\varepsilon}^{*\varepsilon},1\rangle = 0) = 1 - \frac{1}{(1+(t-s-\varepsilon)/\varepsilon)^{1/\beta}}.$$

Then $Q^{*\varepsilon}_{s,\delta_y}\left(e^{-\langle H^{*\varepsilon}_{t-\varepsilon},\phi\rangle} \,|\, \langle H^{*\varepsilon}_{t-\varepsilon},1\rangle > 0\right) = \tilde{u}(s,t-\varepsilon)(y), \quad y \in D^s,$

$0 < s < t-\varepsilon$, where

$$\tilde{u}_{s,t-\varepsilon} = \left(v_{s,t-\varepsilon} - 1 + (1 + \frac{t-s-\varepsilon}{\varepsilon})^{-1/\beta}\right)(1 + \frac{t-s-\varepsilon}{\varepsilon})^{1/\beta}$$

$$\tilde{u}_{t-\varepsilon,t-\varepsilon} = e^{-\phi}.$$

We obtain for $0 \le s \le (t-\varepsilon)$,

$$-\frac{\partial \tilde{u}_{s,t-\varepsilon}}{\partial s} = A_s\tilde{u}_{s,t-\varepsilon} + \frac{1}{\beta(t-s)}\left[(1 - \tilde{u}_{s,t-\varepsilon})^{1+\beta} + \tilde{u}_{s,t-\varepsilon} - 1\right]$$

$$= (A_s - \frac{1}{t-s})\tilde{u}_{s,t-\varepsilon} + \frac{1}{(t-s)\beta}\left[(1 - \tilde{u}_{s,t-\varepsilon})^{1+\beta} + (1+\beta)\tilde{u}_{s,t-\varepsilon} - 1\right].$$

Comparing with (3.27) this completes the proof that the laws of $r_{t-\varepsilon}\tilde{H}^{\varepsilon,x}_t$ and $H^{**}_{t-\varepsilon,t}$ coincide for any $0 < \varepsilon < t$. Next note that for $s < t-\varepsilon$, a.s. $r_s\tilde{H}^{t-s,x}_t(\{y^s\}) = 1$ if and only if $r_s r_{t-\varepsilon}\tilde{H}^{\varepsilon,x}_t(\{y^s\}) \ge 1$ (use (3.18)) and (3.18) implies that the same is true for $H^{**}_{t-\varepsilon,t}$. This implies that the process $\{r_s\tilde{H}^{t-s,x}_t : 0 \le s \le t-\varepsilon\}$ has the same law as $\{H^{**}_{s,t} : 0 \le s \le t-\varepsilon\}$ with initial measure $H^{**}_{0,t} = \delta_x$. Since ε is arbitrary, this completes the proof. ∎

The next result completes the description of H_t in terms of the embedded branching particle systems.

<u>Theorem 3.10.</u> Assume (3.13a), (M_1) and (M_2). Then for fixed $t > 0$,

$$g_b(\varepsilon)\tilde{H}^{\varepsilon}_t \xrightarrow[\varepsilon\downarrow 0]{} H_t, \quad Q_{0,m}\text{-a.s.}$$

<u>Proof.</u> Recall that $\tilde{H}^{\varepsilon}_t \equiv Q_{0,m}(H_t|\mathcal{G}^{*\varepsilon}_t)/g_b(\varepsilon)$ and that the

σ-fields $\mathcal{G}_t^{*\varepsilon} \uparrow$ as $\varepsilon \downarrow 0$. Let $\{\phi_n\}$ be a countable convergence determining family of bounded continuous functions on D^t (Ethier-Kurtz (1986, Ch. 3, (4.5))). By the martingale convergence theorem $g_b(\varepsilon)\langle\tilde{H}_t^\varepsilon,\phi_n\rangle$ converges a.s. as $\varepsilon \downarrow 0$.

In order to complete the proof it suffices to show that $g_b(\varepsilon)\langle\tilde{H}_t^\varepsilon,\phi\rangle \longrightarrow \langle H_t,\phi\rangle$ in probability as $\varepsilon\downarrow 0$, $\phi\in bp\mathcal{D}_t$. It will then follow that $g_b(\varepsilon)\langle\tilde{H}_t^\varepsilon,\phi_n\rangle \longrightarrow \langle H_t,\phi_n\rangle$ $\forall n \in \mathbb{Z}_+$ a.s. and therefore $g_b(\varepsilon)\tilde{H}_t^\varepsilon \longrightarrow H_t$, $Q_{0,m}$-a.s.

Given $\phi_1,\phi_2 \in bp\mathcal{D}_t$ the joint Laplace functional of the pair $(\langle H_t,\phi_1\rangle, g_b(\varepsilon)\langle\tilde{H}_t^\varepsilon,\phi_2\rangle)$ can be obtained by noting that $(H_t, g_b(\varepsilon)\tilde{H}_t^\varepsilon)$ form a Cox cluster random measure with intensity $H_{t-\varepsilon}/g(\varepsilon)$ and pair cluster

$$P^{**}_{t-\varepsilon,t;y}\, e^{-[\langle\mu_1,\phi_1\rangle+\langle\mu_2,\phi_2\rangle]}$$

$$= e^{-(T_{t-\varepsilon,t}\phi_2)(y)\cdot g_b(\varepsilon)}\left(1 - g(\varepsilon)V_{t-\varepsilon,t}\phi_1(y)\right)$$

where we have used Proposition 3.3(b) and (3.10). Therefore

$$\lim_{\varepsilon\downarrow 0} Q_{0,m}\left(e^{-[\langle H_t,\phi_1\rangle+g_b(\varepsilon)\langle\tilde{H}_t^\varepsilon,\phi_2\rangle]}\right)$$

$$= \lim_{\varepsilon\downarrow 0} e^{-\langle m, V_{0,t-\varepsilon}\left\{\frac{1}{g(\varepsilon)}\left[1-P^{**}_{t-\varepsilon,t;y}e^{-[\langle\mu_1,\phi_1\rangle+\langle\mu_2,\phi_2\rangle]}\right]\right\}\rangle}$$

$$= \lim_{\varepsilon\downarrow 0} \exp\Bigg(-\langle m,$$

$$V_{0,t-\varepsilon}\left\{\frac{1}{g(\varepsilon)}\left[1 - e^{-(T_{t-\varepsilon,t}\phi_2)(y)g_b(\varepsilon)}(1-g(\varepsilon)V_{t-\varepsilon,t}\phi_1(y))\right]\right\}\rangle\Bigg)$$

$$= \exp\left(- \lim_{\varepsilon\downarrow 0} [\langle m, V_{0,t-\varepsilon}(T_{t-\varepsilon,t}\phi_2 + V_{t-\varepsilon,t}\phi_1)\rangle]\right)$$

$$\text{(since } g(\varepsilon) \to 0 \text{ and } \lim_{\varepsilon\downarrow 0} \frac{g_b(\varepsilon)}{g(\varepsilon)} = 1)$$

$$= e^{-\langle m, V_{0,t}(\phi_2+\phi_1)\rangle}.$$

The last step follows since $(T_{t-\varepsilon,t}\phi_2 + V_{t-\varepsilon,t}\phi_1) - V_{t-\varepsilon,t}(\phi_1+\phi_2) \xrightarrow{bp} 0$ as $\varepsilon\downarrow 0$ (by (2.1.16)). But this implies that $(\langle H_t,\phi_1\rangle, \langle g_b(\varepsilon)\tilde{H}_t^{\varepsilon},\phi_2\rangle)$ converges in distribution to $(\langle H_t,\phi_1\rangle,\langle H_t,\phi_2\rangle)$ and hence

$$\langle g_b(\varepsilon)\tilde{H}_t^{\varepsilon},\phi_2\rangle - \langle H_t,\phi_2\rangle \longrightarrow 0 \text{ in probability.} \blacksquare$$

Remarks: (1) The usual weak convergence result (e.g. Ethier and Kurtz (1986), Méléard and Roelly-Coppoletta (1989) for a proof in the Feller setting) would state that $g(\varepsilon)H_t^{*\varepsilon}$ converges weakly to H_t as $\varepsilon \to 0$. In this general setting (our state space for W, D(E), is not locally compact and its semigroup may not be Feller) there are additional technical problems. If we assume $T_{s,t}: C(D(E)^t) \longrightarrow C(D(E)^s)$ the techniques in Section 7 show how to overcome this difficulty. In fact for continuous branching the weak convergence is proved without any Feller conditions.

(2) It follows from the weak convergence of the branching particle systems, Theorem 3.9 and Theorem 3.10 that under (SB)

$$d(g(\varepsilon)\tilde{H}^{\varepsilon}_{\cdot}, H_{\cdot}) \longrightarrow 0 \quad \text{in probability as } \varepsilon \downarrow 0,$$

where $d(\cdot,\cdot)$ is a metric for the Skorohod topology on $D(M_F(D(\mathbb{R}^d)))$. (See Lemma A.5 for a proof).

We now specialize to the case of continuous branching

(CB) $\beta = 1, \quad \gamma = \frac{1}{2}$.

The previous results hold for this case but we can obtain a more complete probabilistic description.

Theorem 3.11. Let $0 < s < t$, $m \in M_F(D)^0$ and work with respect to $Q_{0,m}$.

(a) $r_s\tilde{H}_t^{t-s}$ consists of $N_t(s)$ atoms of masses $M_1 = M_1(s)$, $\ldots, M_{N_t(s)} = M_{N_t(s)}(s)$ where $\{N_t(s) : 0 \le s < t\}$ is a time inhomogeneous Markov process with

$$(3.28) \qquad P(N_t(r_2t)=k \,|\, N_t(r_1t)=j) = \binom{k-1}{k-j}\left(\frac{1-r_2}{1-r_1}\right)^j\left(\frac{r_2-r_1}{1-r_1}\right)^{k-j},$$

$$1 \le j \le k, \ k \in \mathbb{Z}_+, \ 0 \le r_1 < r_2 < 1,$$

$$P(N_t(0)=k) = e^{-2\langle m,1\rangle/t}(2\langle m,1\rangle/t)^k/\ k!\ , \ k \in \mathbb{Z}_+.$$

Furthermore, conditioned on $N_t(s) = k$, $M_1(s), \ldots, M_k(s)$ are independent exponentially distributed random variables with mean $(t-s)/2$.

(b) Given $N_t(s) = k$, let τ_i, $i=1,\ldots,k$ denote the splitting times of the particles. Then $\{\tau_i\}$ are i.i.d. and uniformly distributed on $[s,t]$.

(c) The cluster law $P^*_{r,t;y}$ is given by

$$(3.29) \qquad P^*_{r,t;y}(d\nu) = \int \frac{2}{t-r}\, e^{-2\eta/(t-r)} P_{r,t}(\eta, y^r; d\nu)\, d\eta$$

where $P_{r,t}(m, y^r; d\nu)$ denotes the law of the cluster conditioned to have total mass m and is obtained as follows. Starting with one particle of mass m at time r further particles are produced by a processes of subdivision. A particle of mass m_i at time $s < t$ divides with Poisson rate $2m_it/(t-s)^2 ds$ and on splitting produces two particles with masses m_i' and $m_i - m_i'$ and the ratio

m_i'/m_i is uniformly distributed on [0,1].

Proof. (a) Without loss of generality we can assume that $N_t(0) = 1$. Then $M = M_1$, the total mass of the cluster of age t, is exponentially distributed with mean t/2 (cf. (3.7),(3.24) in the case (CB)). From Theorem 3.9(b) $N_t(s)$, $0 \le s \le t$, is equal in law to $\langle H^{**}_{s,t}, 1\rangle$, $0 \le s < t$, and the latter is a time inhomogeneous pure birth process with birth rate at time s equal to $(1/(t-s))$. Then by Theorem 3.6 the generating function $G(s,t;\theta) = E(\theta^{N_t(t-\varepsilon)} | N_t(s)=1)$, $s < t-\varepsilon$, satisfies

$$(3.30) \qquad \frac{-\partial G(s,t-\varepsilon;\theta)}{\partial s} = \frac{G^2-G}{t-s}, \quad G(t-\varepsilon,t-\varepsilon;\theta)=\theta.$$

This yields

$$(3.31) \qquad G(s,t-\varepsilon;\theta) = \frac{\varepsilon\theta/(t-s)}{1 - \theta(t-s-\varepsilon)/(t-s)},$$

that is, geometric with parameter $\varepsilon/(t-s)$. Letting $\varepsilon=t-rt$, $s=0$, we obtain for $0 < r < 1$

$$P(N_t(rt)=k|N_t(0)=1) = r^{k-1}(1-r), \ k \ge 1.$$

In addition for $1 > r_2 > r_1 > 0$, conditioned on $N_t(r_1t)$, $N_t(r_2t)$ is distributed as the sum of $N_t(r_1t)$ geometrics with parameter $[1-(r_2-r_1)/(1-r_1)]$, that is, (3.28) holds.

The fact that $N_t(0)$ is Poisson with mean $2\langle m,1\rangle/t$ follows from Proposition 3.5(b).

Proof of (b) $P(\tau_1 > r(t-s)|N_t(s)=1) = P(N_t(s+r(t-s))=1|N_t(s)=1) = 1-r$, and similarly $P(\min_{i \le k} \tau_i > r(t-s)|N_t(s)=k) = (1-r)^k$, both by (3.28).

Proof of (c). Assume that $N_t(0) = 1$ and let $M = M_1(0)$. We first prove that conditioned $\{M=m\}$, $N_t(rt)-1$ (which represents the number of particle splits by time rt) is an inhomogeneous Poisson process with intensity $\frac{2m}{t(1-r)^2}\,dr$. From (a) M is the sum of $N_t(rt)$ (geometric with mean $1/(1-r)$) masses which are independent exponential (mean $(1-r)t/2$). Thus we obtain the conditional distribution

$$P(N_t(rt)=k|M=m) = \frac{P(N_t(rt)=k).P(M=m|N_t(rt)=k)}{P(M=m)}$$

$$= \frac{(2r)^{k-1}(1-r)\ m^{k-1}e^{-2m/(1-r)t}}{(1-r)^k t^k (k-1)!t^{-1}\ e^{-2m/t}}$$

$$= \left(\frac{2rm}{(1-r)t}\right)^{k-1} \frac{e^{-2mr/(1-r)t}}{(k-1)!}$$

which is Poisson with mean $2rm/(1-r)t$. Similarly for $r_2 > r_1$,

$$P(N_t(r_2t)=k|N_t(r_1t)=j,M=m)$$

$$= \frac{r_1^{j-1}(1-r_1)^{j+1}\binom{k-1}{k-j}\left(\frac{1-r_2}{1-r_1}\right)^j\left(\frac{r_2-r_1}{1-r_1}\right)^{k-j} m^{k-1}e^{-2m/(1-r_2)t}\ t^j(j-1)!}{2^{j-k}\ (1-r_2)^k t^k (k-1)!r_1^{j-1}(1-r_1)\ m^{j-1}e^{-2m/(1-r_1)t}}$$

$$= \frac{(r_2-r_1)^{k-j}\ (2m/t)^{k-j}}{[(1-r_2)(1-r_1)]^{k-j}(k-j)!}\exp\left(-\frac{2m(r_2-r_1)}{t(1-r_1)(1-r_2)}\right)$$

which is Poisson with mean $\left(\dfrac{2(r_2-r_1)m}{(1-r_1)(1-r_2)t}\right)$.

By (b) and (a) the joint density of the time of splitting, τ, of a particle and the masses of the two resulting particles, M_1 and M_2, starting at time $r < t$ is

$$\frac{1}{t-r}\,\frac{4\ e^{-2(m_1+m_2)/(t-s)}}{(t-s)^2} \qquad r \le s < t, \quad m_i > 0.$$

and hence the joint density of τ and $M = M_1+M_2$ is

$$\frac{1}{t-r}\,\frac{4m\ e^{-2m/(t-s)}}{(t-s)^2} \qquad r \le s < t, \quad m > 0.$$

From this we easily obtain that the marginal density of M is exponential with mean $(t-r)/2$,

$$P(\tau > s|M=m) = e^{-m\left[\frac{2}{t-s} - \frac{2}{t-r}\right]} \quad (\text{Poisson rate } 2m/(t-s)^2),$$

and

$$P(M_1 \le \eta M) = \eta, \ \eta \in [0,1].$$

Now consider n particles alive at time $t-r$, conditioned to have masses $M_1,\ldots,M_n$, and let their respective splitting times be $\tau_1,\ldots,\tau_n$. Using (b) and a argument similar to the above we can show that the $\{\tau_i\}$ (and resulting mass ratios) are independent. ∎

Remark: (3.28) was obtained by different methods in Durrett (1978, section 3).

4. Palm Measures and a 0-1 law.

4.1 A Representation for the Historical Palm Measure

In this section we assume, in addition to (H),

(SB) $\Phi((s,y),\lambda) = -\gamma\,\lambda^{1+\beta}$ for some $0 < \beta \le 1$.

$W = (D,\mathcal{D},\mathcal{D}[s,t+],W_t,P_{s,y})$ continues to denote the usual IBSM process in $D^t \subset D$ and $H = (\Omega,\mathcal{G},\mathcal{G}[s,t+],H_t,Q_{s,m})$ is the (W,Φ)-historical process on canonical path space. Let $P_{s,m}(A) = \int P_{s,y}(A)m(dy)$ for $m \in M_F(D)^s$.

We now introduce the Campbell measure and Palm measures for the random measure H_t.

<u>Definition</u> If $s \le t$ and $m \in M_F(D)^s$, the measure on $M_F(D)\times D$ defined by

(4.1.1) $$\bar{Q}(s,m;t,A\times B) = Q_{s,m}(1_A(H_t)H_t(B))$$

is called the Campbell measure associated with H_t. We again choose versions $\{(Q_{s,m,t})_y : y \in D\}$ of the Radon-Nikodym derivatives

(4.1.2) $$\frac{Q_{s,m}(1_A(H_t)H_t(dy))}{Q_{s,m}(H_t(dy))} = \frac{Q_{s,m}(1_A(H_t)H_t(dy))}{P_{s,m}(W_t \in dy)}, \quad A \in \mathcal{B}(M_F(D)),$$

such that $(Q_{s,m,t})_y$ becomes a probability measure on $M_F(D)$ for each $y \in D$ and is Borel measurable in y. Equivalently, $(Q_{s,m,t})_y$ is a regular conditional probability for H_t given y with respect to $\bar{Q}(s,m;t)$. This is possible by a standard argument since $M_F(D)$ is a Polish space. The $(Q_{s,m,t})_y$ are called the Palm distributions associated with the random measure H_t.

We next introduce the Palm measures associated with the canonical measures $R_{s,t}(y,d\nu)$ for H. Recall from Proposition 3.3

that $R_{s,t}(y,.)$ is a finite measure on $M_F(D)$ such that

(4.1.3) $-\log Q_{s,\delta_y}(\exp(-\langle H_t,f\rangle))$

$$= V_{s,t}f(y) = \int(1-e^{-\langle\nu,f\rangle})R_{s,t}(y,d\nu) \quad \forall\ (s,y)\in \hat{E},\ t\geq s,$$

where $V_{s,t}f$ is the unique solution of

(4.1.4) $V_{s,t}f(y) = T_{s,t}f(y) - \gamma\int_s^t T_{s,r}((V_{r,t})^{1+\beta})dr.$

From Proposition 3.3, Lemma 3.4 and (3.24) the total mass of $R_{s,t}(y,.)$ is

(4.1.5) $R_{s,t}(y,M_F(D)) = (\beta\gamma(t-s))^{-1/\beta}$

and the Laplace functional of the total mass of a cluster of age (t-s) is

(4.1.6) $1 - \theta((t-s)\beta\gamma)^{1/\beta}(1+(t-s)\beta\gamma\theta^{\beta})^{-1/\beta}.$

For each $(s,\tilde{y}^s)\in\hat{E}$ and $t\geq s$ let

$$\bar{R}(s,\tilde{y}^s;t,A\times B) = R_{s,t}(\tilde{y}^s,1_A(\mu)\mu(B))$$

denote the Campbell measure associated with $R_{s,t}(y,.)$ on $M_F(D)\times D \ni (H,w)$. Standard arguments give the existence of a jointly measurable regular conditional probability for H given w=y. More precisely for each $s\leq t$ there is a measurable mapping $\tilde{R}_{s,\tilde{y}^s,t}(.)(y)$ from $\{(\tilde{y}^s,y):(\tilde{y}^s,y)\in D^s\times D\}$ to $M_1(M_F(D))$ such that

$$\tilde{R}_{s,\tilde{y}^s,t}(A)(y) = \bar{R}(s,\tilde{y}^s;t,.)(H\in A|w=y), \quad P_{s,\tilde{y}^s}\text{-a.a. } y.$$

We now define the Palm distributions $(R_{s,t})_y$, $(t,y)\in\hat{E}$ associated with $R_{s,t}(y^s,.)$ by setting

$$(R_{s,t})_y(A) = \tilde{R}_{s,y^s,t}(A)(y), \quad A\in\mathcal{B}(M_F(D)).$$

The mapping $y\longrightarrow(R_{s,t})_y$ is a Borel measurable mapping from D^t to $M_1(M_F(D))$. It is easy to check that

$$(R_{s,t})_y(A) = \bar{R}(s,\tilde{y}^s;t,.)(H\in A|w=y), \quad P_{s,\tilde{y}^s}(W_t\in dy)\text{-a.s. } \forall\tilde{y}^s\in D^s.$$

Remark. From the definition of the Palm distributions we easily verify

(4.1.7) $$\iint \langle\psi,\mu\rangle e^{-\langle\phi,\mu\rangle} R_{s,t}(w,d\mu)\,m(dw)$$
$$= \iint \psi(y)\,e^{-\langle\phi,\mu\rangle}(R_{s,t})_y(d\mu)\,P_{s,m}(W_t\in dy) \quad \forall\ \phi,\psi\in bp\mathcal{D}.$$

The relation between the Palm distributions associated to $R_{s,t}$ and H_t, respectively, is given as follows.

Theorem 4.1.1.

$$(Q_{s,m,t})_y(B) = \iint 1_B(\mu_1+\mu_2)\left(q(s,m;t)\times(R_{s,t})_y\right)(d\mu_1 d\mu_2)$$

$$P_{s,m}\text{-a.a. } y\ \forall\ B\in\mathcal{B}(M_F(D))$$

where $q(s,m;t,A) = Q_{s,m}(H_t\in A)$.

Proof. See Kallenberg (1977b, Lemma 10.6) and Liemant et al (1988, 1.8.3).

Now let

(4.1.8) $$Z_{s,t}(\phi,\psi;y^s) = \int \langle\psi,\mu\rangle e^{-\langle\phi,\mu\rangle} R_{s,t}(y,d\mu)$$

$$= \frac{\partial}{\partial\varepsilon}\int\left(1 - e^{-\langle\phi+\varepsilon\psi,\mu\rangle}\right)R_{s,t}(y,d\mu)\Big|_{\varepsilon=0}$$

$$= \frac{\partial}{\partial\varepsilon}\left(V_{s,t}(\phi+\varepsilon\psi)(y)\right)\Big|_{\varepsilon=0} \qquad ,\phi,\psi\in bp\mathcal{D}$$

noting that by (3.15) $\int\langle\psi,\mu\rangle R_{s,t}(y,d\mu) < \infty$, so that the above derivative exists and may be calculated as above.

Then an easy calculation shows that $Z_{s,t}$ satisfies

(4.1.9) $$Z_{s,t}(\phi,\psi) = T_{s,t}\psi - \int_s^t \gamma(1+\beta)T_{s,r}[(V_{r,t}\phi)^\beta Z_{r,t}(\phi,\psi)]dr.$$

Lemma 4.1.2. The solution of (4.1.9) is given by the Feynman-Kac formula

$$(4.1.10) \qquad Z_{s,t}(\phi,\psi;y^s) = P_{s,y^s}\left(e^{-\int_s^t \xi(r,W_r)dr} \psi(W_t)\right)$$

where $\xi(r,.) = (1+\beta)\gamma(V_{r,t}\phi(.))^\beta$

<u>Proof.</u> In view of the uniqueness of solutions to (4.1.9) it suffices to verify that the right side of (4.1.10) satisfies (4.1.9). Substituting the right side of (4.10) into the right side of (4.1.9) we obtain

$$T_{s,t}\psi(y^s) - \int_s^t T_{s,r}\left[\xi(r,.)P_{r,.}\left(e^{-\int_r^t \xi(u,W_u)du} \psi(W_t)\right)\right](y^s)dr$$

$$= T_{s,t}\psi(y^s) - \int_s^t P_{s,y^s}\left(\xi(r,W_r)\ e^{-\int_r^t \xi(u,W_u)du} \psi(W_t)\right)dr$$

by the Markov property

$$= T_{s,t}\psi(y^s) - \int_s^t \frac{\partial}{\partial r} P_{s,y^s}\left\{ e^{-\int_r^t \xi(u,W_u)du} \psi(W_t)\right\}dr$$

(interchange justified since ξ is bounded)

$$= P_{s,y^s}\left\{ e^{-\int_s^t \xi(u,W_u)du} \psi(W_t)\right\}$$

thus proving the result. ∎

<u>Theorem 4.1.3.</u> Let $\phi \in bp\mathcal{D}$. Then for $P_{s,m}$ a.a. y

$$(4.1.11)\quad \int e^{-<\phi,\mu>}(R_{s,t})_{y^t}(d\mu) = e^{-\gamma(1+\beta)\int_s^t (V_{r,t}\phi(y^r))^\beta(y)dr}.$$

<u>Proof.</u> It suffices to show that for $\phi,\psi\in bp\mathcal{D}$,

$$\int\psi(y)\int e^{-<\phi,\mu>}(R_{s,t})_y(d\mu)P_{s,m}(W_t\in dy)$$

$$= \int\psi(y)\left(e^{-\gamma(1+\beta)\int_s^t (V_{r,t}\phi(y^r))^\beta dr} \right)P_{s,m}(W_t\in dy)$$

From (4.1.7) the left hand side is given by

$$\iint<\psi,\mu>e^{-<\phi,\mu>}R_{s,t}(w,d\mu)m(dw).$$

But from (4.1.8)

$$\iint<\psi,\mu>e^{-<\phi,\mu>}R_{s,t}(w,d\mu)m(dw) = \int Z_{s,t}(\phi,\psi;w)m(dw).$$

On the other hand using (4.1.10), the right side becomes

$$= P_{s,m}\left[e^{-\gamma(1+\beta)\int_s^t (V_{r,t}\phi(W_r))^\beta dr}\psi(W_t)\right] = \int Z_{s,t}(\phi,\psi;w)m(dw). \blacksquare$$

<u>Remarks.</u> (1) An analogous expression to (4.1.11) can be obtained for the general Markov case with $\partial\Phi(y,\lambda)/\partial\lambda$ playing the role of $(1+\beta)\lambda^\beta$.

(2) Expression (4.1.11) for the Palm measure is the analogue of the backward tree formula for discrete time branching due to Kallenberg (1977a). An analogous formula for continuous time branching particle systems has recently been obtained by Gorostiza and Wakolbinger (1989).

We now give an alternative probabilistic description of the Palm measures and Campbell measure based on our expression (4.1.10) for the Laplace transform of the Palm measure of the canonical measure $R_{s,t}(y,\cdot)$.

Let $I(\mathbb{R})$ denote the set of non-decreasing paths in $D(\mathbb{R})$ and let $\mathcal{I}$ denote the trace of $\mathcal{D}$ on $I(\mathbb{R})$. Let $M_F^0(D) = M_F(D) - \{0\}$ and

$$\Pi = \{p:[0,\infty) \longrightarrow M_F(D): \{u:p(u) \neq 0\} \text{ is countable}\}.$$

If $p\in \Pi$ and $A \in \mathcal{B}([0,\infty)\times M_F^0(D))$, let

$$N(p)(A) = \sum_{r\geq 0} 1_A(r,p(r)) \in \mathbb{Z}_+\cup\{\infty\}$$ and let $\mathcal{P}$ be the σ-field on Π generated by $\{N(A):A\in \mathcal{B}([0,\infty)\times M_F(D))\}$. If $0\leq s\leq t$, $y\in D(E)$ and $\tau\in I(\mathbb{R})$, let $P_1(s,t,y,\tau)$ denote the unique probability on $(\Pi,\mathcal{P})$ under which $\{N(A):A\in \mathcal{B}([0,\infty)\times M_F^0(D))\}$ is a Poisson point process with characteristic measure

$$\text{(4.1.12)} \quad n(dr,d\nu) = R_{r,t}(y^r,d\nu)1(s\leq r<t)d\tau_r \quad \text{on } [0,\infty)\times M_F^0(D)$$

$$= (\beta\gamma(t-r))^{-1/\beta}P^*_{r,t;y}(d\nu)1(s\leq r<t)d\tau_r$$

(by Proposition 3.3, Lemma 3.4(a) and (3.24)).

(4.1.3) gives the necessary measurability of $R_{r,t}(y^r,A)$ in (r,y^r).

<u>Lemma 4.1.4.</u> If $0\leq s\leq t$ and $A\in \mathcal{P}$, then the mapping $(\tau,y) \longrightarrow P_1(s,t,\tau,y)(A)$ is $\mathcal{I}\times\mathcal{D}$-measurable.

<u>Proof.</u> If $f:[0,\infty)\times M_F^0(D) \longrightarrow [0,\infty)$, then

$$\text{(4.1.13)} \quad P_1(s,t,\tau,y)\left(\exp\left(-\iint f(r,\nu)N(dr,d\nu)\right)\right)$$

$$\exp\left\{-\int 1(s\leq r<t)\int_{M_F^0(D)} (1 - e^{-f(r,\nu)})R_{r,t}(y^r,d\nu)d\tau_r\right\}.$$

The right side is $\mathcal{I}\times\mathcal{D}$-measurable by (4.1.3). A monotone class argument completes the proof. ∎

Let P_1 be the law on $(I(\mathbb{R}),\mathcal{I})$ of the stable subordinator, T_t, of index $\beta\in(0,1]$ (β as in (SB)), starting at zero and scaled so that

$$P_1(e^{-\theta T_t}) = \exp\{-\gamma(1+\beta)\theta^\beta t\}$$

or equivalently

$$(4.1.14)\quad P_1\left(\exp\left(-\int f(r)dT(r)\right)\right) = \exp\left\{-\gamma(1+\beta)\int f(t)^\beta dt\right\},$$
$$f\in bp\mathcal{B}([0,\infty)).$$

Define $P_1(s,t,y) = P_1(s,t,y^t)$ on $(\Pi,\mathcal{P})$ by

$$P_1(s,t,y)(A) = \int P_1(s,t,y,\tau)(A)P_1(d\tau).$$

<u>Proposition 4.1.5.</u> For all $0\le s\le t$ and $m\in M_F(D)^s$,

$$P_1(s,t,y)\left(\iint \nu N(dr,d\nu)\in A\right) = (R_{s,t})_{y^t}(A) \quad \forall\ A\in\mathcal{B}(M_F(D))$$
$$\text{for } P_{s,m}\text{-a.a. } y.$$

<u>Proof.</u> If $\phi\in bp\mathcal{D}$,

$$P_1(s,t,y)\left(\exp\left\{-\iint\langle\nu,\phi\rangle N(dr,d\nu)\right\}\right)$$
$$= P_1\left(\exp\left\{-\int_s^t\int_{M_F^0(D)}(1-e^{-\langle\nu,\phi\rangle})R_{r,t}(y^r,d\nu)dT(r)\right\}\right)$$
$$\text{(see (4.1.13))}$$
$$= P_1\left(\exp\left\{-\int_s^t V_{r,t}\phi(y^r)dT(r)\right\}\right) \quad \text{(by (4.1.3))}$$
$$= \exp\left\{-\gamma(1+\beta)\int_s^t (V_{r,t}\phi(y^r))^\beta dr\right\} \quad \text{(by (4.1.14))}$$
$$= \int e^{-\langle\nu,\phi\rangle}(R_{s,t})_{y^t}(d\nu)$$

$P_{s,m}$-a.a. y (by (4.1.11)). The above equality holds for all $\phi\in bp\mathcal{D}$ for $P_{s,m}$-a.a. y (consider a countable collection of ϕ's in $bp\mathcal{D}$ whose bounded pointwise closure is $bp\mathcal{D}$). The result now follows from the uniqueness of the Laplace functional. ∎

Corollary 4.1.6. Let $\beta=1$, $\gamma = 1/2$.

(a) For $\eta \in \mathbb{R}^+$, $y \in D$ let $P_{r,t}(\eta, y^r; d\nu)$ denote the probability law of a cluster of age $t-r$ conditioned to have mass η, and starting from $y(r)$ at time r as constructed in Theorem 3.11(c).

Let $\{N(r): s \le r < t\}$ be an inhomogeneous Poisson point process on $[0,\infty)\times M_F(D)$ with Lévy measure

$$n(dr, d\eta, d\nu) = \frac{2}{t-r}\left[\frac{2}{t-r}\, e^{-2\eta/(t-r)}\right] 1(s \le r < t)\; P_{r,t}(\eta, y^r; d\nu)\, dr\, d\eta$$

Then $P\left(\iiint \nu\, N(dr, d\eta, d\nu)) \in A\right) = (R_{s,t})_y(A)$.

(b) $$(R_{s,t})_y(e^{-\theta<1,\nu>}) = \left(\frac{2/(t-s)}{\theta + 2/(t-s)}\right)^2.$$

Proof. (a) See (3.29), (4.1.5), (4.1.12) and use the previous result.

(b) $$(R_{s,t})_y(e^{-\theta<1,\nu>}) = \exp\left\{-\int_s^t\int (1-e^{-\theta x}) \left(\frac{2}{t-r}\right)^2 e^{-2x/(t-r)} dx dr\right\}$$

$$= \left(\frac{2/(t-s)}{\theta + 2/(t-s)}\right)^2. \;\blacksquare$$

Remark. In the case $\beta=1$, the total mass of a cluster of age t is exponential with mean $t/2$, since (see (3.29))

$$P^*_{s,t;y}(\{\nu: <\nu,1> \in dx\}) = ((t-s)/2)^{-1} \exp\{-2x/(t-s)\} dx .$$

On the other hand, the total mass of a Palm cluster of age t has probability density

$$(4/t^2)x\ e^{-2x/t}, \quad x\geq 0.$$

Thus the Palm cluster has a mean twice that of an ordinary cluster. This reflects the fact that the Palm measure favours larger clusters.

If $0\leq s<t$ and $m\in M_F(D)^S$ define a finite measure, $\bar{P}_1 = \bar{P}_1(s,m;t)$ on $(\bar{\Omega}_1,\bar{\mathcal{F}}_1) = (\Pi\times D,\mathcal{P}\times\mathcal{D})$ by

$$\bar{P}_1(A\times B) = \int 1_B(y)P_1(s,t,y)(A)P_{s,m}(dy)$$

and set

$$(\bar{\Omega},\bar{\mathcal{F}},\bar{P}=\bar{P}(s,m;t)) = (\bar{\Omega}_1\times M_F(D),\bar{\mathcal{F}}_1\times\mathcal{B}(M_F(D)),\bar{P}_1(s,m;t)\times q(s,m;t))$$

where $q(s,m;t) = Q_{s,m}(H_t\in\cdot)$

Points in $\bar{\Omega}$ are denoted by $\omega = (p,y,\tilde{H})$ and we will abuse the notation and consider the point process N as a random measure on $(\bar{\Omega},\bar{\mathcal{F}})$ or $(\bar{\Omega}_1,\bar{\mathcal{F}}_1)$. The random measure which will interest us, is defined on $(\bar{\Omega},\bar{\mathcal{F}})$ by

$$(4.1.15)\quad H = \tilde{H} + \iint \nu N(dr,d\nu)$$

$$\text{i.e.}\quad H(A) = \tilde{H}(A) + \iint \nu(A)N(dr,d\nu),\quad A\in\mathcal{D}.$$

Using (4.1.15) one sees that (recall $D = D(E)$)

$$\int_D\int_\Pi\iint \nu(D)N(dr,d\nu)dP_1(s,t,\tau,y)(p)dP_{s,m}(y)$$

$$= \int_D\int_{[s,t)}\int_{M_F(D)} \nu(D)R_{r,t}(y^r,d\nu)d\tau_r dP_{s,m}(y)$$

$$= m(D)(\tau_t - \tau_s) < \infty \quad \text{(see (3.15))}.$$

It follows that

(4.1.16) $\int_{[s,t)} \nu N(dr,d\nu) \in M_F(D)$ and hence $H \in M_F(D)$,

$\bar{P}(s,m;t)$-a.s.

<u>Theorem 4.1.7.</u> (a)

$$P_1(s,t,y)\times q(s,m;t)\left(\tilde{H} + \iint \nu N(dr,d\nu))\in A\right) = (Q_{s,m,t})_y(A),$$

$\forall\ A\in\mathcal{B}(M_F(D))$ for $P_{s,m}(W_t\in dy)$-a.a.y.

(b) $\bar{P}(s,m;t)((H,y)\in A\times B) = \bar{Q}(s,m;t,A\times B)$ for all $A\in\mathcal{B}(M_F(D))$, $B\in\mathcal{D}$, $t>s\geq 0$ and $m\in M_F(D)^s$.

<u>Proof.</u> (a) This follows from Proposition 4.1.5 and Theorem 4.1.1.

(b) By definition

$$\bar{Q}(s,m;t,A\times B) = Q_{s,m}(1_A(H_t)H_t(B))$$

$$= \int 1_B(y)(Q_{s,m,t})_y(A)P_{s,m}(W_t\in dy)$$

$$= \int 1_B(y)(P_1(s,t,y)\times q(s,m,t))\left(\ \tilde{H} + \iint \nu N(dr,d\nu) \in A\right)P_{s,m}(W_t\in dy)$$

(by (a))

$$= (\bar{P}_1(s,m;t)(.\times B)\times q(s,m,t))\left(\ \tilde{H} + \iint \nu N(dr,d\nu) \in A\right)$$

$$= \bar{P}(s,m;t)((H,y)\in A\times B). \blacksquare$$

<u>Remark.</u> $\bar{Q}(0,m;t,A\times D)/m(D)$ is the law of H_t conditioned on non-extinction in the future. This is shown in Evans-Perkins (1989) (the local compactness of the state space assumed there is not needed for the above result). See also Roelly-Coppoletta and Rouault (1989) where the Laplace functional of $X_t = \bar{\Pi}_t(H_t)$ (X conditioned on non-extinction) is also computed (in our setting this is immediate from Theorem 4.1.3). Theorem 4.1.7 "explains" the product form of the Laplace functional. The "immigration term" (see the discussion at the end of Roelly-Coppoletta and

Rouault (1989)) arises from the cousins of the tagged trajectory y which has now been averaged with respect to $P_{0,m}(W_t\in\cdot)$.

We thank the referee for pointing out the above remark. The next result helps motivate the decomposition of H.

Notation. If $y\in D(E)$, let

$$y^{t-}(u) = \left\{ \begin{array}{ll} y(u) & \text{if } u<t \\ y(t-) & \text{if } u\geq t \end{array} \right\} \in D(E).$$

Proposition 4.1.8. Let $s\geq 0$ and assume (3.18).

(a) $\displaystyle\int_{[r,t)}\int_{M_F(D)} \nu(A)N(du,d\nu) = H(\{w\in A: w^{r-}=y^{r-}\})\ \forall A\in\mathcal{D}, r\in(s,t),$

$\bar{P}(s,m;t)$-a.s. $\forall t>s$, $m\in M_F(D)^S$.

(b) For $t > 0$, H_t is non-atomic $Q_{0,m}$-a.s.

Proof. (a) This follows by a first moment argument and we omit it.

(b) Let $g: D\times M_F(D) \longrightarrow \mathbb{R}^+$ be jointly measurable. Then from the definition of Palm measures it follows that

$$\iint g(y,\mu)\mu(dy)Q_{0,m}(H_t\in d\mu) = \iint g(y,\mu)(Q_{0,m,t})_y(d\mu)Q_{0,m}(H_t(dy)).$$

Letting $g(y,\mu) = \mu(\{y\})$ it suffices to show that

$$\int e^{-\mu(\{y^t\})}(Q_{0,m,t})_{y^t}(d\mu) = 1,\ P_{0,m}\text{-a.a. } y.$$

Fix $y\in D$ and let $\phi_\varepsilon(w) \downarrow 1_{\{y^t\}}(w)$ as $\varepsilon\downarrow 0$. By Theorem 4.1.1,

$$\int e^{-\mu(\{y^t\})}(Q_{0,m,t})_{y^t}(d\mu)$$

$$\geq \lim_{\varepsilon\downarrow 0}\int e^{-\langle\phi_\varepsilon,\mu\rangle}Q_{0,m}(d\mu)\cdot\int e^{-\langle\phi_\varepsilon,\mu\rangle}(R_{0,t})_{y^t}(d\mu)$$

$$= \lim_{\varepsilon\downarrow 0}\int e^{-\langle\phi_\varepsilon,\mu\rangle}Q_{0,m}(d\mu)\cdot\exp\left\{-\gamma(1+\beta)\int_0^t (V_{r,t}\phi_\varepsilon(y^r))^\beta dr\right\}$$

(by (4.1.11))

$= 1$

since $Q_{0,m}(\langle\phi_\varepsilon,H_t\rangle) \downarrow 0$ and for $r < t$, $V_{r,t}\phi_\varepsilon(y^r) \overset{bp}{\longrightarrow} 0$ as $\varepsilon \downarrow 0$ under (3.18). Hence $\mu(\{y^t\}) = 0$, $(Q_{0,m,t})_{y^t}$ -a.s. and the proof is complete. ∎

Remarks. (1) The same proof as for (b) shows that if $P^x(Y_t=y) = 0$ for all x,y, then the (Φ,Y)-superprocess X_t has no atoms a.s.

(2) From the viewpoint of the particle picture one sees that y^t denotes the trajectory of a tagged particle up to time t and (from the above result) $\int_{[s,t)}\int_{M_F(D)} \nu N(du,d\nu)$ denotes the contribution to H_t from those cousins of y which branched off at time $u\in[s,t)$. The component $\tilde{H}$ of the decomposition (4.1.15) represents the contribution to H_t from particles which are unrelated to y. The ancestral decomposition (4.1.15) was used (in the particle setting) in many of the arguments in Perkins (1988).

4.2 A 0-1 Law

Notation. $\mathscr{S}_x(x') = x'-x$, $x',x \in \mathbb{R}^d$. If $0\le u\le t$ and $(y,\tau)\in D(\mathbb{R}^d)\times I(\mathbb{R})$ let

$$\tilde{Y}^t_u(y,\tau) = (y(t)-y((t-u)-),\tau(t)-\tau((t-u)-)) = (\tilde{y}^t(u),\tilde{\tau}^t(u))$$

$$\tilde{\mathcal{D}}^t_u = \sigma(\tilde{Y}^t_s : s\le u).$$

Recall if $h: E_1\longrightarrow E_2$ is a measurable mapping between metric spaces E_i then $\bar{h}:M_F(E_1) \longrightarrow \bar{h}(M_F(E_2))$ is defined by $\langle\bar{h}(\nu),f\rangle = \langle\nu,f\circ h\rangle$, $f \in b\mathcal{B}(E_2)$.

If $E = \mathbb{R}^d$, $0\le u\le v<\infty$ and $(p,y) \in \bar{\Omega}_1$, define random measures on $(\mathbb{R}^d,\mathcal{B}(\mathbb{R}^d))$ by

$$X_{u,v}(p,y) = \mathcal{G}_{y(t)} \circ \bar{\Pi}_t\left(\int_{[u,v)}\int_{M_F(D)} \nu \, N(p)(dr,d\nu)\right)$$

$$X_t^u(p,y) = X_{t-u,t}(p,y), \quad 0 \le u \le t.$$

We may, and shall, consider $X_{u,v}$ and X_t^u as random measures on $(\bar{\Omega},\bar{\mathcal{F}})$ as well as $(\bar{\Omega}_1,\bar{\mathcal{F}}_1)$. Define sub-$\sigma$-fields of $\bar{\mathcal{F}}$ by

$$\mathcal{H}_t^\delta = \sigma(X_t^u(A): 0<u\le\delta,\ A\in \mathcal{B}(\mathbb{R}^d)), \quad \mathcal{H}_t^{0+} = \bigcap_{\delta>0} \mathcal{H}_t^\delta.$$

It follows easily from (4.1.16) that the mapping $u \longrightarrow X_t^u$ is in $D([0,t],M_F(\mathbb{R}^d))$ $\bar{P}(s,m;t)$ a.s. for all $s<t$, $m\in M_F(D)^s$.

In this section we assume (SB) and

(L) $Y = (D,\mathcal{D},\mathcal{D}_{t+},Y_t,P^x)$ is a Lévy process in $\mathbb{R}^d$, defined on canonical path space and with semigroup P_t.

H is the (Y,Φ)-historical process (and hence W and $P_{s,y}$ are constructed from Y as in Theorem 2.2.1).

An immediate consequence of (L) is

(4.2.1) $P_1\times P_{s,m}(\tilde{Y}^t(.\wedge(t-s))\in A)/m(D)$

$$= P_1\times P^o((T(.\wedge(t-s)),Y(.\wedge(t-s)))\in A) \quad \forall\ A\in \mathcal{I}\times\mathcal{D},$$

$0\le s<t$ and $m\in M_F(D)^s$.

In this Markovian setting, nothing is gained by considering $s>0$. We identify $D(\mathbb{R}^d)^0$ with $\mathbb{R}^d$ and for $m\in M_F(\mathbb{R}^d)$ let

$$\bar{P}(m;t) = \bar{P}(0,m;t), \quad \bar{P}_1(m;t) = \bar{P}_1(0,m;t),$$

$$P_m = P_{0,m} = \int P^x m(dx) \equiv P^m, \quad P_1(t,\tau,y) = P_1(0,t,\tau,y),$$

$$P_1(t,y) = P_1(0,t,y), \quad \bar{Q}(m;t,A) = \bar{Q}(0,m;t,A).$$

<u>Lemma 4.2.1.</u> If $f\in b\mathcal{B}(D(M_F(\mathbb{R}^d)))$, $0\le u\le t$ and $m\in M_F(\mathbb{R}^d)$, then

(a) $(y,\tau) \longrightarrow \int_\Pi f(X_t^{\cdot\wedge u}(p,y))P_1(t,\tau,y)(dp)$ is $\tilde{\mathcal{D}}_u^t$-measurable.

(b) The $\bar{P}(m;t)m(D)^{-1}$-distribution of $X_t^{\cdot\wedge u}$ depends only on the choice of u and not on the choice of m or t satisfying $t \ge u$.

<u>Proof.</u> Choose $0=u_0<u_1<\dots<u_n\le u$ and $f_i \in bp\mathcal{B}(\mathbb{R}^d)$ $(i \le n)$, and let $U_r f_i$ denote the unique solution (see Theorem 2.1.3) of

$$(4.2.2)\quad U_r f_i(x) = P_r f_i(x) - \gamma\int_0^r P_u(U_{r-u}(\cdot)^{1+\beta})(x)\,du.$$

If $g_j(x) = \sum_{i=1}^{j} f_i(x)$, then

$$\sum_{i=1}^{n} \langle X_t^{u_i}, f_i\rangle = \sum_{j=1}^{n} \langle X_{t-u_j,\,t-u_{j-1}}, g_j\rangle$$

and so, if $y \in D(\mathbb{R}^d)$ and we consider $X_t^{u_i}(\cdot,y)$ as a random measure defined on $(\Pi,\mathcal{P})$ we have

$$P_1(t,\tau,y)\Big(\exp\Big\{-\sum_{i=1}^{n}\langle X_t^{u_i}, f_i\rangle\Big\}\Big)$$

$$= \exp\Big\{-\sum_{j=1}^{n}\iint 1(t-u_j \le r < t-u_{j-1})$$

$$(1-\exp\{-\langle \nu, g_j\circ\mathcal{S}_{y(t)}\circ\Pi_t\rangle\})R_{r,t}(y^r,d\nu)\,d\tau_r\Big\}$$

(by (4.1.13))

$$= \exp\Big\{\sum_{j=1}^{n}\int 1(t-u_j \le r < t-u_{j-1}) \log Q_{r,\delta_{y^r}}(\exp\{-\langle H_t, g_j\circ\mathcal{S}_{y(t)}\circ\Pi_t\rangle\})\,d\tau_r\Big\}$$

(by (4.1.3))

$$(4.2.3)$$

$$= \exp\Big\{\sum_{j=1}^{n}\int 1(t-u_j \le r < t-u_{j-1}) \log Q_{0,\delta_{y(r)}}(\exp\{-\langle X_{t-r}, g_j\circ\mathcal{S}_{y(t)}\rangle\})\,d\tau_r\Big\}$$

by Theorem 2.2.4, where X is the (Y,Φ)-superprocess starting at m under Q_m. By Theorem 2.1.3, (4.2.3) equals

$$(4.2.4) \quad \exp\left\{-\sum_{j=1}^{n}\int 1(t-u_j\le r<t-u_{j-1})\, U_{t-r}(g_j\circ\mathcal{S}_{y(t)})(y(r))\,d\tau_r\right\}.$$

Since $P_r(f\circ\mathcal{S}_x) = (P_rf)\circ\mathcal{S}_x$, the uniqueness of (4.2.2) shows that $U_r(f\circ\mathcal{S}_x) = (U_rf)\circ\mathcal{S}_x$. Substitute this into (4.2.4) to conclude

$$P_1(t,\tau,y)\left(\exp\left\{-\sum_{i=1}^{n}\langle X_t^{u_i},f_i\rangle\right\}\right)$$

$$= \exp\left\{-\sum_{j=1}^{n}\int 1(t-u_j\le r<t-u_{j-1})\, U_{t-r}g_j(y(r)-y(t))\,d\tau_r\right\}$$

$$(4.2.5) \quad = \exp\left\{-\sum_{j=1}^{n}\int 1(u_{j-1}<r\le u_j)\, U_rg_j(-\tilde{y}^t(r-))\,d\tilde{\tau}(r)\right\}.$$

The right side is $\tilde{\mathcal{D}}_u^t$-measurable and (a) follows by a standard monotone class argument.

For (b), integrate out (τ,y) in (4.2.5) and conclude

$$\bar{P}(m;t)\left(\exp\left\{-\sum_{i=1}^{n}\langle X_t^{u_i},f_i\rangle\right\}\right)m(\mathbb{R}^d)^{-1}$$

$$= P_1\times P^m\left(\exp\left\{-\sum_{j=1}^{n}\int_{(u_{j-1},u_j]} U_rg_j(-\tilde{y}^t(r-))\,d\tilde{\tau}^t(r)\right\}\right)m(\mathbb{R}^d)^{-1}$$

$$= P_1\times P^o\left(\exp\left\{-\sum_{j=1}^{n}\int_{(u_{j-1},u_j]} U_rg_j(-Y(r-))\,dT(r)\right\}\right) \quad \text{(by (4.2.1))}.$$

The right side is independent of the choice of t or $m\in M_F(\mathbb{R}^d)$, providing $t\ge u$. Hence the finite-dimensional distributions of $X_t^{\cdot\wedge u}$ are independent of the choice of t or m and (b) follows. ∎

<u>Notation.</u> Let $X_u(\omega) = \omega(u)$ denote the coordinate mappings on $D(M_F(\mathbb{R}^d))$, $\mathcal{M}_u = \sigma(X_r: r\le u)$ and let $\bar{\mathcal{M}}_{0+}$ denote the universal completion of $\mathcal{M}_{0+}$ relative to $\mathcal{B}(D(M_F(\mathbb{R}^d)))$. More precisely, if

$\mathcal{N}(P)$ is the class of P-null sets in $D(M_F(\mathbb{R}^d))$, then $\bar{\mathcal{M}}_{0+} = \bigcap_P(\mathcal{M}_{0+} \vee \mathcal{N}(P))$ where the intersection is over all probabilities P on $\mathcal{B}(D(M_F(\mathbb{R}^d)))$.

The notation X_t is, strictly speaking, in conflict with the notation $X_t = \bar{\Pi}_t(H_t)$ from Section 2 but since the only probabilities on $D(M_F(\mathbb{R}^d))$ we will study are the laws Q_m of (Y,Φ)-superprocesses, this will not create any ambiguity (recall Theorem 2.2.4).

<u>Theorem</u> <u>4.2.2.</u> If $h:D(M_F(\mathbb{R}^d)) \longrightarrow [-\infty,\infty]$ is $\bar{\mathcal{M}}_{0+}$-measurable, then there is a constant $c = c(h) \in [-\infty,\infty]$ such that

$$(4.2.6) \quad h(X_t^{\cdot\wedge t}) = c \quad \bar{P}(m;t)\text{-a.s. } \forall\ t>0 \text{ and } m\in M_F(\mathbb{R}^d).$$

In particular,

$$(4.2.7) \quad \bar{P}(m;t)(A) = 0 \quad \text{or} \quad \bar{P}(m;t)(A^c) = 0 \quad \forall\ A\in \mathcal{H}_t^{0+},\ m\in M_F(\mathbb{R}^d),\ t>0.$$

<u>Proof.</u> As the factor $\tilde{H}$ is independent of $X_t^{\cdot}$ and the σ-fields $\mathcal{H}_t^{0+}$ we may, and shall, work on $(\bar{\Omega}_1,\bar{\mathcal{F}}_1)$ with $\bar{P}_1(m;t)$ in place of $\bar{P}(m;t)$. Hence we are considering X_t^u as a random measure on $(\bar{\Omega}_1,\bar{\mathcal{F}}_1)$ and $\mathcal{H}_t^{0+}$ as a sub-σ-field of $\bar{\mathcal{F}}_1$.

Fix $t>0$ and $m\in M_F(\mathbb{R}^d)$ and consider (4.2.7) first. Let

$$\mathcal{H}[r,v) = \sigma(X_{r',v}:r\le r'<v) \subset \bar{\mathcal{F}}_1.$$

Clearly $\lim_{v\uparrow t} X_{t-u,v} = X_{t-u,t} = X_t^u$ and hence

$$\mathcal{H}_t^{\delta} = \mathcal{H}[t-\delta,t) = \bigvee_{n=1}^{\infty} \mathcal{H}[t-\delta,t-2^{-n}).$$

Let $A \in \mathcal{H}_t^{0+}$. By the above there are $m_n>n$ and $A_n \in$

$\mathcal{H}[t-2^{-n},t-2^{-m_n})$ such that $\bar{P}_1(m;t)(A\Delta A_n) < 2^{-n}$. Choose B_n, C_n in $\mathcal{B}(D(M_F(\mathbb{R}^d)))$ such that

$$A_n = \{X_{u_n(\cdot),t-2^{-m_n}} \in B_n\}, \quad A = \{X_t^{\cdot\wedge 2^{-m_n}} \in C_n\}$$

where $u_n(u) = t - (u\wedge 2^{-n})\vee 2^{-m_n}$. Then

$$\bar{P}_1(m;t)(A\cap A_n)$$

$$= \int_I\int_D P_1(t,\tau,y)(X_{u_n(\cdot),t-2^{-m_n}} \in B_n, X_t^{\cdot\wedge 2^{-m_n}} \in C_n)P_m(dy)P_1(d\tau)$$

(4.2.8) $$= \int_D P_1(t,\tau,y)(X_{u_n(\cdot),t-2^{-m_n}} \in B_n)P_1(t,\tau,y)(X_t^{\cdot\wedge 2^{-m_n}} \in C_n) P_m(dy)P_1(d\tau)$$

because $X_{u_n(\cdot),t-2^{-m_n}}$ and $X_t^{\cdot\wedge 2^{-m_n}}$ are independent under $P_1(t,\tau,y)$ since they correspond to integrals w.r.t. N over disjoint time intervals. Note that

(4.2.9) $$P_1(t,\tau,y)(\{p:X_t^{\cdot\wedge 2^{-m_n}}(p,y) \in C_n\})$$
$$= P_1(t,\tau,y)(\{p:(p,y) \in A\}).$$

The left side is $\tilde{\mathcal{D}}^t_{2^{-m_n}}$-measurable (Lemma 4.2.1) and the right side is independent of n. The σ-field $\tilde{\mathcal{D}}^t_{0+}$ is $P_1\times P_m$-trivial (see (4.2.1)) and so (4.2.9) is $P_1\times P_m$-a.s. equal to a constant, $c = c(m,t,A)$. Integrate out the right hand side of (4.2.9) w.r.t. $P_1\times P_m$ to see that

(4.2.10) $$\bar{P}_1(m;t)(A) = c(m,t,A)m(\mathbb{R}^d).$$

Substitute this constant into (4.2.8) to get

$$\bar{P}_1(m;t)(A \cap A_n) = c(m,t,A)\bar{P}_1(m;t)(A_n)$$

and then let $n \longrightarrow \infty$ and use (4.2.10) to see that

$$c = \bar{P}_1(m;t)(A)m(\mathbb{R}^d)^{-1} = c(m,t,A)\bar{P}_1(m;t)(A)m(\mathbb{R}^d)^{-1} = c^2.$$

Hence $c(m,t,A) = 0$ or 1 and (4.2.7) is proved.

Turning to (4.2.6), let h be as in the Theorem and (m,t) be as above. Choose a second (m_1,t_1) and write $(m,t) = (m_2,t_2)$ for notational convenience. If $P(A) = \sum_{i=1}^{2} \bar{P}_1(m_i;t)(X_{t_i}^{\cdot \wedge t_i} \in A)$ then $h \in \mathcal{M}_{0+} \vee \mathcal{N}(P)$ and so there is a $g \in \mathcal{M}_{0+}$ such that

(4.2.11) $$h(X_{t_i}^{\cdot \wedge t_i}) = g(X_{t_i}^{\cdot \wedge t_i}) \quad \bar{P}_1(m_i;t_i)\text{-a.s.}, \ i=1,2.$$

$g(X_{t_i}^{\cdot \wedge t_i}) \in \mathcal{H}_{t_i}^{0+}$ and so by (4.2.7) there are constants $c_i = c(m_i,t_i) \in [-\infty,\infty]$ such that

$$g(X_{t_i}^{\cdot \wedge t_i}) = c_i \quad \bar{P}_1(m_i;t_i)\text{-a.s.}$$

g is $\mathcal{M}_{0+}$-measurable and therefore

(4.2.12) $$c_i = \lim_{u \downarrow 0} g(X_{t_i}^{\cdot \wedge u}) \quad \bar{P}_1(m_i;t_i)\text{-a.s.}$$

Lemma 4.2.1(b) shows that the $\bar{P}_1(m_i;t_i)m_i(\mathbb{R}^d)^{-1}$-law of the right side of (4.2.12) is independent of (m_i,t_i) and so $c_1=c_2=c$. Returning to (4.2.11) with $i=2$ we obtain

$$h(X_t^{\cdot \wedge t}) = c \quad \bar{P}_1(m;t)\text{-a.s.}$$

Since $c = c(m_1,t_1)$ does not depend on (m,t), the proof of (4.2.6) is complete. ∎

<u>Remark</u> <u>4.2.3.</u> Although we believe $\bar{\Omega}$ (or $\bar{\Omega}_1$) is the proper

setting for the above "0-1" law, it may at times be convenient to have a version on the canonical space $M_F(D)\times D$ for $\bar{Q}(s,m;t,.)$. This is possible by Proposition 4.1.8 if we assume (3.18) or equivalently in our Lévy setting

(4.2.13) $P^0\times P^0(Y_s^1 = Y_s^2 \ \forall\ s\le u) = 0 \ \forall\ u>0.$

If $\omega = (H,y) \in M_F(D)\times D$ and $0\le u\le t$, let

$$\tilde{H}_t^u(\omega)(A) = H(\{w\in A: w^{(t-u)-} = y^{(t-u)-}\}),\ A \in \mathcal{D}$$

$$\tilde{X}_t^u(\omega) = \mathcal{G}_{y(t)}\circ\bar{\Pi}_t(\tilde{H}_t^u(\omega))$$

$$\tilde{\mathcal{H}}_t^\delta = \sigma(\tilde{X}_t^u: 0<u\le\delta),\quad \tilde{\mathcal{H}}_t^{0+} = \bigcap_{\delta>0} \tilde{\mathcal{H}}_t^\delta.$$

Theorem 4.2.2 and Proposition 4.1.8 imply that, under (4.2.13), (SB) and (L), if $h: D(M_F(\mathbb{R}^d)) \longrightarrow [-\infty,\infty]$ is $\bar{\mathcal{M}}_{0+}$-measurable there is a $c\in [-\infty,\infty]$ such that

$$h(\tilde{X}_t^{\cdot\wedge t}) = c \quad \bar{Q}(m;t)\text{-a.s.} \quad \forall \quad t>0,\ m\in M_F(\mathbb{R}^d)$$

and in particular,

$$\bar{Q}(m;t,A) = 0 \quad \text{or} \quad \bar{Q}(m;t,A^c) = 0 \quad \forall\ A\in \tilde{\mathcal{H}}_t^{0+},\ m\in M_F(\mathbb{R}^d),\ t>0.$$

5. Hausdorff Measure and the Support of Super-Levy Processes

In this section we use the "0-1" law of the previous section (Theorem 4.2.2) to refine the results on the supports of super-Lévy processes obtained in Perkins (1988,1989) and Dawson-Iscoe-Perkins (1989). We work with continuous branching in the Lévy setting and hence assume

(CB) $\Phi((s,y),\lambda) = -\lambda^2/2$

and (L) (see the beginning of 4.2). For the most part we work on the spaces $(\bar{\Omega},\bar{\mathcal{F}},\bar{P}(m;t))$, $m\in M_F(\mathbb{R}^d)$, $t>0$, constructed in Section 4. Let $\bar{X}_t = \bar{\Pi}_t(H)$ and $\tilde{X}_t = \bar{\Pi}_t(\tilde{H})$ where H is given by (4.1.15) on $(\bar{\Omega},\bar{\mathcal{F}})$. Theorems 4.1.7 and 2.2.4 imply that for any $m\in M_F(\mathbb{R}^d)$, $t>0$, $A\in \mathcal{M} = \mathcal{B}(M_F(\mathbb{R}^d))$ and $B\in \mathcal{B}(\mathbb{R}^d)$,

(5.1) $\bar{P}(m;t)(\tilde{X}_t\in A, Y_t\in B) = Q_m(X_t\in A)P^m(Y_t\in B)$,

(5.2) $\bar{P}(m;t)(\bar{X}_t\in A, Y_t\in B) = Q_m(1_A(X_t)X_t(B))$,

where Q_m is the law of the $(Y,-\lambda^2/2)$-super process on $C(M_F(\mathbb{R}^d))$. Therefore

(5.3) $\bar{P}(m;t)(\bar{X}_t\in .) << Q_m(X_t\in .)$ on $M_F(\mathbb{R}^d)$,

<u>Notation.</u> $\mathcal{I} = \{\psi:[0,\varepsilon] \to [0,\infty): \varepsilon>0,\ \psi$ continuous and strictly increasing, $\psi(0)=0,\ \underline{\lim}_{n\to\infty} \psi(2^{-n-1})/\psi(2^{-n}) > 0\}$. $B(x,r)$ denotes the closed ball in $\mathbb{R}^d$ with centre x and radius r.

In order to apply the "0-1" law, Theorem 4.2.2, we first show that for small r we can ignore the contribution to $\bar{X}_t(B(y_t,r))$ from cousins which branched off from y_t before $t-r^\beta$, where $\beta < 2-4/d$.

Lemma 5.1. Assume Y is a d-dimensional Brownian motion, $d\geq 3$. If $0 < \beta < 2-4/d$, $m\in M_F(\mathbb{R}^d)$ and $t>0$, then $\bar{P}(m;t)$-a.s. $\exists$ $r_0=r_0(\beta,t,\omega) > 0$ such that

$$\bar{X}_t(B(Y_t,r)) - X_t^{r^\beta}(B(0,r))$$

$$= \tilde{X}_t(B(Y_t,r)) + \int_{[0,t-r^\beta)}\int \bar{\Pi}_t(\nu)(B(Y(t),r))N(ds,d\nu) = 0,$$

$\forall\ 0<r\leq r_0$.

Proof. The first equality is immediate from the definitions. Fix β,m,t as above with $m\neq 0$. Theorem 3.1(a) of [DIP] implies

$$(5.4)\qquad Q_m(X_t(B(x,r)) > 0) \leq c_1(d)r^{d-2}\ t^{-d/2}m(\mathbb{R}^d)\quad \forall\ t\geq r^2.$$

Therefore (use (5.1))

$$(5.5)\qquad \bar{P}(m;t)(\tilde{X}_t(B(Y_t,r)) > 0) \leq c_1(d)r^{d-2}\ t^{-d/2}m(\mathbb{R}^d)^2 \quad \forall\ t\geq r^2.$$

Also if $y\in C = C(\mathbb{R}^d)$, $r\in (0,1]$ and $u\in [0,t-r^\beta]$

$$\exp\left\{-\int_{M_F(C)} (1-e^{-\lambda\bar{\Pi}_t(\nu)(B(y_t,r))})R_{u,t}(y^u,d\nu)\right\}$$

$$= Q_{u,\delta_{y^u}}(\exp\{-\lambda\ \bar{\Pi}_t(H_t)(B(y_t,r))\}) \qquad \text{(by (3.9)}$$

$$= Q_{\delta_{y(u)}}(\exp\{-\lambda X_{t-u}(B(y_t,r))\}) \qquad \text{(Theorem 2.2.4)}$$

$$\longrightarrow Q_{\delta_{y(u)}}(X_{t-u}(B(y_t,r))=0) \quad \text{as } \lambda\longrightarrow\infty$$

$$(5.6)\qquad \geq 1 - c_1(d)r^{d-2}(t-u)^{-d/2} \qquad \text{by (5.4)}$$

because $t-u\geq r^\beta\geq r^2$. Choose $r_1=r_1(d,\beta)$ sufficiently small positive such that $c_1(d)r_1^{d-2}\ r_1^{-d\beta/2}\leq 1/2$. If $r\in (0,r_1)$, then

$$\bar{P}(t,m)\left(\int_{[0,t-r^\beta)}\int \bar{\Pi}_t(\nu)(B(Y_t,r))N(ds,d\nu) > 0\right)m(\mathbb{R}^d)^{-1}$$

$$= 1 - \lim_{\lambda\to\infty} \bar{P}(t,m)\Big(\exp\Big\{-\lambda \int_{[0,t-r^\beta)}\int \bar{\Pi}_t(\nu)(B(Y_t,r))N(ds,d\nu)\}\big)m(\mathbb{R}^d)^{-1}$$

$$= 1 - \lim_{\lambda\to\infty} P^m\Big(\exp\Big\{-\int_0^{t-r^\beta}\int(1-\exp\{-\lambda\ \bar{\Pi}_t(\nu)(B(Y_t,r))\})R_{u,t}(Y^u,d\nu)du\Big\}\Big)$$

(see (4.1.13))

$$\le 1 - \exp\Big\{-\int_0^{t-r^\beta} \log\Big(1/(1-c_1(d)r^{d-2}(t-u)^{-d/2})\Big)du\Big\} \quad \text{(by (5.6))}$$

$$\le \int_0^{t-r^\beta} \log\Big(1/(1-c_1(d)\ r^{d-2}(t-u)^{-d/2})\Big)du \quad (1-e^{-x} \le x \quad \forall\ x\ge 0)$$

$$\le t\ \log(1/(1-c_1(d)r^{d-2}\ r^{-d\beta/2}))$$

$$\text{(5.7)} \qquad \le t(2\log\ 2)c_1(d)\ r^{d-2-d\beta/2},$$

where in the last line we have used the choice of r_1 to conclude $c_1(d)r^{d-2-d\beta/2} \le 1/2$ and the fact that $\log 1/(1-\varepsilon) \le (2\log 2)\varepsilon$ for $\varepsilon \in (0,1/2]$. The upper estimates (5.5) and (5.7) allow the obvious Borel-Cantelli argument to complete the proof. ∎

Recall from the Introduction that $\psi_\alpha(r) = r^\alpha \log\log \frac{1}{r}$ and $S(X_t)$ is the closed support of X_t.

<u>Theorem 5.2.</u> If X_t is super Brownian motion in $\mathbb{R}^d$, $d\ge 3$, starting at $m\in M_F(\mathbb{R}^d)$ under Q_m, then there is a $c(d)\in(0,\infty)$ such that

$$X_t(A) = c_d\ \psi_2\text{-}m(A\cap S(X_t)) \quad \forall\ A\in \mathcal{B}(\mathbb{R}^d) \quad Q_m\text{-a.s.} \quad \forall\ t > 0,$$
$$m\in M_F(\mathbb{R}^d).$$

<u>Proof.</u> Fix $t > 0$ and $m\in M_F(\mathbb{R}^d)$. Define $\tilde{S}:M_F(\mathbb{R}^d) \longrightarrow M(\mathbb{R}^d)$ by $\tilde{S}(\nu)(A) = \psi_2\text{-}m(S(\nu)\cap A)$. By Perkins (1989) Q_m-a.s. X_t and $\tilde{S}(X_t)$ are equivalent measures and there are universal constants $0 <$

$c_1(d) \le c_2(d) < \infty$ such that

$$(5.8)\qquad c_1(d) \le \frac{dX_t}{d\tilde{S}(X_t)}(x) \equiv f(x,X_t) \le c_2(d), \quad X_t\text{-a.a. } x \quad Q_m\text{-a.s.}$$

By the Besicovitch-Morse Theorem (cf. Federer (1969, p. 16)) we may define a version of $f(x,X_t)$ by

$$f(x,X_t) = \lim_{\substack{r\downarrow 0\\ r\in Q}} \frac{X_t(B(x,r))}{\tilde{S}(X_t)(B(x,r))}$$

where the limit exists for X_t a.a. x Q_m-a.s. Define $f(x,X_t) \equiv 0$ on the exceptional set, N, of (x,X_t) where the limit does not exist. Note therefore that

$$\begin{aligned} f(x,\nu) &= \lim_{\substack{r\downarrow 0\\ r\in Q}} \frac{\nu(B(x,r))}{\tilde{S}(\nu)(B(x,r))} \mathbf{1}_{N^c}(x,\nu) \\ &= \lim_{\substack{r\downarrow 0\\ r\in Q}} \frac{\bar{\mathcal{T}}_x(\nu)(B(0,r))}{\tilde{S}(\bar{\mathcal{T}}_x(\nu))(B(0,r))} \mathbf{1}_{N^c}(x,\nu) \end{aligned}$$

where

$$N^c = \{(x,\nu): \lim_{\substack{r\downarrow 0\\ r\in Q}} \bar{\mathcal{T}}_x(\nu)(B(0,r))/\tilde{S}(\bar{\mathcal{T}}_x(\nu))(B(0,r)) \text{ exists}\}$$

satisfies

$$(5.9)\qquad \int \mathbf{1}_N(x,X_t)X_t(dx) = 0 \quad Q_m\text{-a.s.}$$

Lemma 6.3 of [DIP] implies that $f(x,\nu)$ is Borel measurable on $\mathbb{R}^d \times M_F(\mathbb{R}^d)$ (note that $(x,\nu) \longrightarrow \bar{\mathcal{T}}_x(\nu)$ is clearly Borel measurable).

If $0 < \beta < 2 - 4/d$, then

$$f(y_t,\bar{X}_t) = \lim_{\substack{r\downarrow 0\\ r\in Q}} \frac{\bar{X}_t(B(y_t,r))}{\tilde{S}(\bar{X}_t)(B(y_t,r))} \quad \bar{P}(m;t)\text{-a.s.} \text{ (by (5.2) and}$$

(5.9))

$$\text{(5.10)} \qquad = \lim_{\substack{r\downarrow 0\\ r\in Q}} \frac{X_t^{r^\beta}(B(0,r))}{\tilde{S}(X_t^{r^\beta})(B(0,r))} \qquad \bar{P}(m;t)\text{-a.s.} \quad \text{(by Lemma 5.1)}.$$

Let $g(X_t^{\cdot\wedge t})$ be defined by (5.10) if this limit exists and set it equal to 0 elsewhere. It is easy to see from Lemma 6.3 of [DIP] that g is $\bar{\mathcal{M}}_{0+}$-measurable. Theorem 4.2.2 implies there is a constant $c = c(d) \in [0,\infty]$, independent of (t,m) and therefore depending only on d (set $\beta = 1 - 2/d$) such that

$$f(y_t, \bar{X}_t) = g(X_t^{\cdot\wedge t}) = c(d) \quad \bar{P}(m;t)\text{-a.s.}$$

By (5.2) this implies (recall $f(x,\nu)$ is Borel measurable)

$$f(x,X_t) = c(d) \quad X_t\text{-a.a. } x \quad Q_m\text{-a.s.}$$

(5.8) shows that $c(d) \in (0,\infty)$. The definition of $f(x,X_t)$ shows $X_t(A) = c(d)\psi_2(A\cap S(X_t)) \quad \forall\, A\in \mathcal{B}(\mathbb{R}^d)$, Q_m-a.s. ∎

Let $F(\mathbb{R}^d)$ denote the set of closed subsets of $\mathbb{R}^d$ with the Hausdorff metric topology. Recall that if

$$d(x,A) = \inf_{y\in A} |x-y|\wedge 1, \qquad x\in \mathbb{R}^d, \quad A \subset \mathbb{R}^d,$$

the Hausdorff metric on $F(\mathbb{R}^d)$ is defined by

$$\rho(A,B) = \Big(\sup_{x\in A} d(x,B)\Big) \vee \Big(\sup_{x\in B} d(x,A)\Big).$$

Here $\inf \emptyset = 1$ and $\sup \emptyset = 0$, so that $\emptyset$ is a discrete point in $F(\mathbb{R}^d)$ (see Cutler (1984, Ch. 4)). By Theorem 4.4.1 of Cutler (1984) the support mapping $S: M_F(\mathbb{R}^d) \longrightarrow F(\mathbb{R}^d)$ is Borel measurable. Let $K(\mathbb{R}^d)$ denote the space of compact subsets of $\mathbb{R}^d$. Then $K(\mathbb{R}^d)$ with the subspace topology is a locally compact separable metric space (Dugundji (1966, p. 253)).

For our next result we continue to consider the law, Q_m, of super-Brownian motion on $\mathcal{B}(D(M_F(\mathbb{R}^d))$ (which is supported by $C(M_F(\mathbb{R}^d))$. If $S_t = S(X_t)$ is the closed support of super-Brownian motion, then (Perkins (1990, Proposition 4.2)), $\{S_t:t>0\}$ (respectively, $\{S_t:t\geq 0\}$) has càdlàg paths in $K(\mathbb{R}^d)$ Q_m-a.s. for any m in $M_F(\mathbb{R}^d)$ (respectively, for any m in $M_F(\mathbb{R}^d)$ with compact support). Recall $\mathcal{M}_t = \sigma(X_u:u\leq t)$. Let

$$S_t^o = \sigma(S(X_u):u\leq t); \quad \bar{\mathcal{M}}_t^m = \mathcal{M}_t \vee \{Q_m\text{-null sets in } \mathcal{B}(D(M_F(\mathbb{R}^d)))\},$$
$$\bar{S}_t^m = S_t^o \vee \{Q_m\text{-null sets in } \mathcal{B}(D(M_F(\mathbb{R}^d)))\}.$$

It is natural to ask (and Rick Durrett asked it) whether or not $S(X_t)$ is itself a $K(\mathbb{R}^d)$-valued Markov process under Q_m. As mentioned in the Introduction, this follows easily from Theorem 5.2.

<u>Theorem</u> <u>5.3.</u> Let Q_m denote the law of super-Brownian motion in $\mathbb{R}^d$ with $d\geq 3$.

(a) $\bar{\mathcal{M}}_t^m = \bar{S}_t^m \quad \forall\ t\geq 0,\ m\in M_F(\mathbb{R}^d)$.

(b) If $\psi \in b\mathcal{B}(K(\mathbb{R}^d))$, $t\geq 0$ and $u>0$, then

$$Q_m(\psi(S_{t+u})\,|\,\bar{S}_t^m) = Q_m(\psi(S_{t+u})\,|\,S_t) \quad Q_m\text{-a.s.}$$

<u>Proof.</u> (a) The inclusion $\bar{S}_t^m \subset \bar{\mathcal{M}}_t^m$ is clear from the Borel measurability of the support mapping discussed above. For $v\leq t$ fixed, Theorem 5.2 shows that $X_v = c\psi_2\text{-}m(.\cap S_v)$ Q_m-a.s. and so $X_v \in \bar{S}_t^m$ (see [DIP, Lemma 6.3]). This gives the converse inclusion.

(b) If ψ, t and u are as above, then

$$Q_m(\psi(S_{t+u})\,|\,\bar{S}_t^m) = Q_m(\psi\circ S(X_{t+u})\,|\,\bar{\mathcal{M}}_t^m) \quad \text{(by (a))}$$

$$= Q_{X_t}(\psi\circ S(X_u)) \quad Q_m\text{-a.s.},$$

by the Markov property for X. This last expression is $\overline{\sigma(S_t)}^m$-measurable by Theorem 5.2, where $\overline{}^m$ again indicates the inclusion of all Q_m-null sets. Therefore

$$Q_m(\psi(S_{t+u})|\bar{S}_t^m) = Q_m(\psi(S_{t+u})|\overline{\sigma(S_t)}^m) \quad Q_m\text{-a.s.}$$
$$= Q_m(\psi(S_{t+u})|\sigma(S_t)) \quad Q_m\text{-a.s.} \quad \blacksquare$$

Remark 5.4. Our discussion of the $K(\mathbb{R}^d)$-valued Markov process ends here because the strong Markov property of S_t remains unresolved. A glance at the above argument shows that it suffices to find a means of recovering X_t from its closed support which is valid for all $t>0$ Q_m-a.s. For example, if one could establish the conclusion of Theorem 5.2 for all t simultaneously Q_m-a.s., the strong Markov property would follow. Our suspicion, and hope, is that an easier proof of the strong Markov property will be found.

As was mentioned in the Introduction, if $Y = Y_\alpha$ is symmetric α-stable in $\mathbb{R}^d$ with $\alpha<2$, (see the Introduction for the scaling of Y_α) then $S(X_t) = \mathbb{R}^d$ or $\emptyset$ a.s. and Theorem 5.2 no longer holds. We must replace $S(X_t)$ by a Borel support consisting of points of positive ψ_α-density for X_t.
We introduce some notation to state the precise result:
If $\nu \in M_F(\mathbb{R}^d)$, $\psi:[0,\varepsilon] \longrightarrow [0,\infty)$ is non-decreasing, let

$$\Gamma(\nu,\psi,c) = \{x: \overline{\lim_{r\downarrow 0}}\ \nu(B(x,r))/\psi(r) \geq c\}$$

$$\Lambda(\nu,\psi,c) = \{x: \overline{\lim_{r\downarrow 0}}\ \nu(B(x,r))/\psi(r) = c\}$$

$$\Gamma_-(\nu,\psi,c) = \{x: \overline{\lim_{r\downarrow 0}}\ \nu(B(x,r))/\psi(r) \leq c\}$$

Recall that by a Theorem of Rogers and Taylor (1961) (see Perkins (1988, Theorem 1.4)) there is a constant $c_0(d)$ such that

(5.11) $\quad \nu(A\cap\Gamma(\nu,\psi,c)) \geq c_0(d)c\ \psi\text{-}m(A\cap\Gamma(\nu,\psi,c)) \quad \forall\ A\in \mathcal{B}(\mathbb{R}^d)$,

(5.12) $\quad \nu(A\cap\Gamma_-(\nu,\psi,c)) \leq c\ \psi\text{-}m(A\cap\Gamma_-(\nu,\psi,c)) \quad \forall\ A\in \mathcal{B}(\mathbb{R}^d)$.

<u>Theorem 5.5.</u> If $d > \alpha$, there are constants $c_1(\alpha,d) \in (0,\infty)$ and $c_2(\alpha,d) \in [c_0(d)c_1(\alpha,d), c_1(\alpha,d)]$ such that if X is the $(Y_\alpha, -\lambda^2/2)$-superprocess starting at $X_0 = m \in M_F(\mathbb{R}^d)$ under Q_m, then for any $t > 0$

(a) $\overline{\lim\limits_{r\downarrow 0}}\ X_t(B(x,r))/\psi_\alpha(r) = c_1(\alpha,d) \qquad X_t\text{-a.a. } x, \quad Q_m\text{-a.s.}$

(b) $X_t(A) = c_2(\alpha,d)\psi_\alpha\text{-}m(A\cap\Lambda(X_t,\psi_\alpha,c_1(\alpha,d)))$
$= c_2(\alpha,d)\psi_\alpha\text{-}m(A\cap\Gamma(X_t,\psi_\alpha,c_1(\alpha,d))) \quad \forall\ A\in \mathcal{B}(\mathbb{R}^d) \quad Q_m\text{-a.s.}$

The next result will play the role of Lemma 5.1 in the proof of Theorem 5.5, which will be given at the end of this section.

<u>Lemma 5.6.</u> Let $\psi\in \mathcal{I}$. In addition to (L) assume $\exists\ \gamma>1$ such that

(5.13) $$\lim_{r\downarrow 0} \sup_{\substack{\delta\le t\le \delta^{-1}\\ x\in\mathbb{R}^d}} P^0\times P^0(Y_t^1 - Y_t^2 \in B(x,r))(\log 1/r)^\gamma \psi(r)^{-1} = 0 \qquad \forall\delta>0,$$

where Y^1 and Y^2 denote independent copies of Y starting at 0. Then $\forall m\in M_F(\mathbb{R}^d)$ and $t>0$,

(5.14) $$\lim_{r\downarrow 0} (\bar{X}_t(B(y_t,r)) - X_t^u(B(0,r)))\psi(r)^{-1} = 0 \qquad \forall u\in [0,t]$$

$\bar{P}(m;t)$-a.s.

<u>Remark.</u> (a) Since $\mathcal{I}_{y(t)}(\bar{X}_t)-X_t^u$ is a measure, clearly $\bar{X}_t(B(y_t,r))-X_t^u(B(0,r)) \ge 0$.

(b) The condition $\gamma>1$ is by no means critical. The proof is an elementary first moment calculation and this suffices for particular cases of interest, namely when Y is symmetric α-stable in $\mathbb{R}^d$ with $d>\alpha$. More delicate results may be derived by using higher moments. These would be needed if we were to consider the critical case $d = \alpha$ where the conjectured exact

Hausdorff measure function is $\psi(r) = r^{\alpha}(\log \frac{1}{r})(\log\log\log \frac{1}{r})$ (see Perkins (1988, (1.6)). Unfortunately we do not yet know if (1.4) holds (with ψ in place of ψ_α) when $d = \alpha$ and hence have contented ourselves with the above simple formulation of Lemma 5.4.

<u>Proof of Lemma 5.6.</u> If $m \in M_F(\mathbb{R}^d)$, $t>0$, $\delta \in (0,t]$ and $r > 0$, then

$$\bar{P}(m;t)\left(\sup_{u\in[\delta,t]} |\bar{X}_t(B(Y_t,r)) - X_t^u(B(0,r))|\right)$$

$$= \bar{P}(m;t)\left(\tilde{X}_t(B(Y_t,r)) + \int_{[0,t-\delta)}\int_{M_F(\mathbb{R}^d)} \bar{\Pi}_t(\nu)(B(y(t),r))N(ds,d\nu)\right)$$

$$= Q_m \times P^m(X_t(B(Y_t,r)) + P^m\left(\int_0^{t-\delta}\int \bar{\Pi}_t(\nu)(B(Y_t,r))R_{u,t}(Y^u,d\nu)du\right)$$

(by (5.1) and (4.1.12))

$$= P^m\times P^m(|Y_t^1-Y_t^2|\le r) + \int_0^{t-\delta} P^m(Q_{u,\delta_{Y^u}}(\bar{\Pi}_t(H_t)(B(Y_t,r)))du$$

(by (3.15) and (2.1.7) where Y^1, Y^2 are i.i.d. copies of Y)

$$= P^m\times P^m(|Y_t^1-Y_t^2|\le r) + \int_0^{t-\delta}\int P^{Y_u(y)}(|Y'_{t-u}-Y_t(y)|\le r)P^m(dy)du$$

(Theorem 2.2.4 and (2.1.7); here Y' is a copy of Y starting at $Y_u(y)$)

$$= P^m\times P^m(|Y_t^1-Y_t^2|\le r) + \int_0^{t-\delta} P^0\times P^0(|Y_{t-u}^1-Y_{t-u}^2|\le r)du\ m(\mathbb{R}^d)$$

$$\le m(\mathbb{R}^d)^2 c(\delta,t,r) +' m(\mathbb{R}^d)t\ c(\delta,t,r),$$

where $c(\delta,t,r) = \sup_{\substack{\delta\le s\le t\\ x\in\mathbb{R}^d}} P^0\times P^0(Y_s^1-Y_s^2 \in B(x,r))$. (5.13) implies that for small r, $c(\delta,t,r) \le \psi(r)(\log 1/r)^{-\gamma}$ and hence for small r

$$(5.15)\quad \bar{P}(m;t)\Big(\sup_{u\in[\delta,t]} |\bar{X}_t(B(y_t,r)) - X_t^u(B(0,r))|\psi(r)^{-1}\Big)$$

$$\le m(\mathbb{R}^d)(m(\mathbb{R}^d)+t)(\log 1/r)^{-\gamma}.$$

The right side of (5.15) is summable if we set $r = 2^{-n}$. Use the fact that $r \longrightarrow \bar{X}_t(B(y_t,r)) - X_t^u(B(0,r)) \ge 0$ is non-decreasing in r and $\psi(2^{-n-1})/\psi(2^{-n}) \ge \delta > 0$ for large n, to see that since

$$\lim_{n\to\infty} \sup_{u\in[\delta,t]} |\bar{X}_t(B(y_t,2^{-n}))-X_t^u(B(0,2^{-n}))|\psi(2^{-n})^{-1} = 0, \ \bar{P}(m;t)\text{-a.s.}$$

(by Borel-Cantelli), then in fact

$$\lim_{r\downarrow 0} \sup_{u\in[\delta,t]} |\bar{X}_t(B(y_t,r))-X_t^u(B(0,r))|\psi(r)^{-1} = 0, \quad \bar{P}(m;t)\text{-a.s.} \ \blacksquare$$

<u>Theorem 5.7.</u> Assume (L) and (5.13). Let $\psi\in \mathcal{I}$.

(a) $\exists\ c_1\in [0,\infty]$ such that $\forall t>0$, $m\in M_F(\mathbb{R}^d)$

$$\overline{\lim_{r\downarrow 0}}\ X_t(B(x,r))/\psi(r) = c_1, \quad X_t\text{-a.a. } x \quad Q_m\text{-a.s.}$$

(b) If $c_1\in (0,\infty)$, $\exists\ c_2\in [c_0(d)c_1,c_1]$ ($c_0(d)$ as in (5.11)) such that $\forall t > 0$ and $m\in M_F(\mathbb{R}^d)$

$$X_t(A) = c_2\psi\text{-}m(A\cap\Lambda(X_t,\psi,c_1)) = c_2\psi\text{-}m(A\cap\Gamma(X_t,\psi,c_1)) \quad \forall A\in\mathcal{B}(\mathbb{R}^d) \ Q_m\text{-a.s.}$$

<u>Proof.</u> (a) If $m\in M_F(\mathbb{R}^d)$ and $t > 0$, Lemma 5.6 implies

$$\overline{\lim_{r\downarrow 0}}\ X_t(B(y_t,r))\psi(r)^{-1} = h(X_t^{\cdot\wedge t}), \quad \bar{P}(m;t)\text{-a.s.}$$

where $h(x) = \overline{\lim_{u\downarrow 0}}\ \overline{\lim_{r\downarrow 0}}\ x(u)(B(0,r))\psi(r)^{-1}$ is clearly $\mathcal{M}_{0+}$-measurable on $D(M_F(\mathbb{R}^d))$. Theorem 4.2.2 implies there is a $c_1 \in [0,\infty]$, independent of (t,m) such that

$$\overline{\lim_{r\downarrow 0}}\ X_t(B(y_t,r))\psi(r)^{-1} = c_1 \quad \bar{P}(m;t)\text{-a.s.}$$

(5.2) completes the derivation of (a).

(b) Let $t > 0$, $m \in M_F(\mathbb{R}^d)$. Define $\tilde{\Gamma}: M_F(\mathbb{R}^d) \longrightarrow M(\mathbb{R}^d)$ by $\tilde{\Gamma}(\nu)(A) \equiv \psi\text{-}m(A \cap \Gamma(\nu,\psi,c_1))$. (5.11), (5.12) and (a) imply that Q_m a.s.

$$(5.16)\quad c_0(d)c_1\tilde{\Gamma}(X_t)(A) \le X_t(A \cap \Gamma(X_t,\psi,c_1)) = X_t(A)$$

$$= X_t(A \cap \Lambda(X_t,\psi,c_1)) \le c_1\tilde{\Gamma}(X_t)(A) \ \forall\ A \in \mathcal{B}(\mathbb{R}^d).$$

As in the proof of Theorem 5.2, if

$$f(x,\nu) = \begin{cases} \lim\limits_{r\downarrow 0, r\in Q} \nu(B(x,r))/\tilde{\Gamma}(\nu)(B(x,r)) & \text{if } (x,\nu) \in N^c = \{(x,\nu): \text{this limit } \exists\} \\ 0 & \text{otherwise,} \end{cases}$$

then

$$\int 1_N(x,X_t)X_t(dx) = 0, \quad Q_m\text{-a.s.},$$

f is universally measurable on $\mathbb{R}^d \times M_F(\mathbb{R}^d)$ (see Corollary A.4), and

$$(5.17)\quad \frac{dX_t}{d\tilde{\Gamma}_t(X_t)}(x) = f(x,X_t) \in [c_0(d)c_1, c_1] \quad X_t\text{-a.a. } x \quad Q_m\text{-a.s.}$$

As before we will use (5.2) to work on $(\bar{\Omega}, \bar{\mathcal{F}}, \bar{P}(m;t))$. Let

$$\bar{X}_t^u = \int_{[t-u,t]}\int_{M_F(D)} \bar{\Pi}_t(\nu)N(ds,d\nu) = \bar{\mathcal{G}}_{-y(t)}(X_t^u), \quad 0 < u \le t.$$

Assume $w = (p,y,\tilde{H})$ is chosen outside a $\bar{P}(m;t)$-null set so that (5.14) holds, as well as the following conditions:

$$(5.18)\quad (y_t, \bar{X}_t) \in N^c$$

$$(5.19)\quad \overline{\lim_{r\downarrow 0, r\in Q}}\ \bar{X}_t(B(y_t,r))/\psi(r) = c_1$$

$$(5.20)\quad \bar{X}_t(\Lambda(\bar{X}_t,\psi,c_1)^c) = 0$$

(use (5.2), (5.3) and (a) - the necessary measurability for (5.20) is readily established). As in the derivation of (5.16), (5.20) implies

$$(5.21)\quad c_0(d)c_1\tilde{\Gamma}(\bar{X}_t)(A) \le \bar{X}_t(A) \le c_1\Gamma(X_t)(A) \quad \forall\ A \in \mathcal{B}(\mathbb{R}^d).$$

Let $\varepsilon \in (0,c_1)$ and $u \in (0,t]$. If $r \in Q$ satisfies

(5.22) $X_t^u(B(0,r)) > c_1\psi(r)/2,$

then

$$|\psi\text{-}m(\Gamma(X_t^u,\psi,c_1-\varepsilon)\cap B(0,r))/\tilde{\Gamma}(\bar{X}_t)(B(y_t,r))-1|$$
$$\le [\psi\text{-}m((\Gamma(\bar{X}_t,\psi,c_1)-\Gamma(\bar{X}_t^u,\psi,c_1-\varepsilon))\cap B(y_t,r))$$
$$+ \psi\text{-}m((\Gamma(\bar{X}_t^u,\psi,c_1-\varepsilon)-\Gamma(\bar{X}_t,\psi,c_1))\cap B(y_t,r))]\tilde{\Gamma}(\bar{X}_t)(B(y_t,r))^{-1}$$
$$\le [\psi\text{-}m(\Gamma(\bar{X}_t-\bar{X}_t^u\ ,\psi,\varepsilon)\cap B(y_t,r))$$
$$+\psi\text{-}m(\Gamma(\bar{X}_t,\psi,c_1-\varepsilon)\cap\Gamma(\bar{X}_t,\psi,c_1)^c)]c_1{}^{-1}\bar{X}_t(B(y_t,r))^{-1}$$

(by (5.21)

$$(5.23)\quad \le c_0(d)^{-1}\varepsilon^{-1}(\bar{X}_t(B(y_t,r))-X_t^u(B(0,r)))2\psi(r)^{-1}$$
$$+ c_0(d)^{-1}(c_1-\varepsilon)^{-1}\bar{X}_t(\Gamma(\bar{X}_t,\psi,c_1-\varepsilon)\cap\Gamma(\bar{X}_t,\psi,c_1)^c)2\psi(r)^{-1}$$

(by (5.11) amd (5.22)).

The second term in (5.23) is zero by (5.20) and the first term goes to zero as $r\downarrow 0$ by (5.14). (5.14) and (5.19) show that (5.22) holds for arbitrarily small positive rational r $\forall$ $u\in(0,t]$. We have proved (for our fixed choice of ω)

$$(5.24)\quad \lim_{r\downarrow 0}{}^{*u}\ \psi\text{-}m(\Gamma(X_t^u,\psi,c_1-\varepsilon)\cap B(0,r))/\ \tilde{\Gamma}(\bar{X}_t)(B(y_t,r)) = 1$$

$\forall$ $u\in(0,t]$,

where $\lim\limits_{r\downarrow 0}{}^{*u}$ means that the limit is taken over rationals $r > 0$ satisfying (5.22). Note also that $\forall u\in(0,t]$

$$\overline{\lim_{r\downarrow 0}}{}^{*u}\ (\bar{X}_t-\bar{X}_t^u)(B(y_t,r))/\ \tilde{\Gamma}(\bar{X}_t)(B(y_t,r))$$

$$\le \overline{\lim_{r\downarrow 0}}{}^{*u}(\bar{X}_t(B(y_t,r))-X_t^u(B(0,r)))2/\psi(r) \quad \text{(by (5.21) and (5.22))}$$

$$(5.25)\quad = 0 \qquad \text{(by (5.14))}.$$

Since $(y_t,\bar{X}_t)\in N^c$ ((5.18)), we have for any $u\in(0,t]$,

$$f(y_t,\bar{X}_t) = \lim_{r\downarrow 0}{}^{*u}\ \bar{X}_t(B(y_t,r))/\ \tilde{\Gamma}(\bar{X}_t)(B(y_t,r))$$

$$= \lim_{r\downarrow 0}{}^{*u}\ (X_t^u(B(0,r))/\ \psi\text{-}m(\Gamma(X_t^u,\psi,c_1-\varepsilon)\cap B(0,r))$$

$$\times\psi\text{-}m(\Gamma(X_t^u,\psi,c_1-\varepsilon)\cap B(0,r))/\ \tilde{\Gamma}(\bar{X}_t)(B(y_t,r)))$$

$$+\ (\bar{X}_t-\bar{X}_t^u)(B(y_t,r))/\ \tilde{\Gamma}(\bar{X}_t)(B(y_t,r))$$

$$(5.26)\qquad = \lim_{r\downarrow 0}{}^{*u}\ X_t^u(B(0,r))/\ \psi\text{-}m(\Gamma(X_t^u,\psi,c_1-\varepsilon)\cap B(0,r))$$

by (5.24) and (5.25)).

Define $g:D(M_F(\mathbb{R}^d)) \to [0,\infty]$ by

$$g(x) = \overline{\lim_{k\to\infty}}\ \lim_{n\to\infty}\ \sup\Big\{x(k^{-1})(B(0,r))/\ \psi\text{-}m(\Gamma(x(k^{-1}),\psi,c_1-\varepsilon)\cap B(0,r)):$$

$$r\in (0,n^{-1})\cap Q,\ x(k^{-1})(B(0,r)) > c_1\psi(r)/2\Big\}$$

($\sup\emptyset = 0$). g is universally measurable by Corollary A.4 in the appendix. Since $g(x) = g(x(.\wedge n^{-1}))$ $\forall$ $n\in\mathbb{N}$, it is easy to show that g is $\bar{\mathcal{M}}_{0+}$-measurable. By (5.26) and Theorem 4.2.2 there is a constant $c_2\in [0,\infty]$, independent of (t,m), such that

$$f(y_t,\bar{X}_t) = g(X_t^{\cdot\wedge t})\qquad \bar{P}(m;t)\text{-a.s.}$$

$$= c_2\qquad \bar{P}(m;t)\text{-a.s.}$$

By (5.2) this is equivalent to

$$f(x,X_t) = c_2\quad X_t\text{-a.a.}x\quad Q_m\text{-a.s.}$$

(5.17) shows that $c_2\in [c_0(d)c_1,c_1]$ and

$$(5.27)\quad X_t(A) = c_2\psi\text{-}m(A\cap\Gamma(X_t,\psi,c_1))\quad \forall\ A\in \mathcal{B}(\mathbb{R}^d)\quad Q_m\text{-a.s.}$$

Finally (5.11) shows that

$$\psi\text{-}m((\Gamma-\Lambda)(X_t,\psi,c_1)) \le c_0(d)^{-1}c_1{}^{-1}X_t((\Gamma-\Lambda)(X_t,\psi,c_1)) = 0\quad Q_m\text{-a.s.}$$

(by (a)).

Hence we may replace Γ with Λ on the right side of (5.27) and the proof is complete. ∎

<u>Proof of Theorem 5.5.</u> If Y_t^1 and Y_t^2 are independent symmetric α-stable processes in $\mathbb{R}^d$ which start at 0, then

$$\sup_{\substack{\delta\le t\le\delta^{-1}\\ x\in\mathbb{R}^d}} P(Y^1(t)-Y^2(t) \in B(x,r))(\log 1/r)^2\psi_\alpha(r)^{-1}$$
$$\le c(\delta)r^{d-\alpha}(\log 1/r)^2(\log\log 1/r)^{-1} \longrightarrow 0 \quad \text{as} \quad r \to 0+.$$

Hence (5.13) holds and Theorem 5.7(a) proves Theorem 5.5(a) but with $c_1(\alpha,d) \in [0,\infty]$. Theorem 6.5 of Perkins (1988) shows $c_1(\alpha,d) \in (0,\infty)$. This then allows us to deduce Theorem 5.5(b) from Theorem 5.7(b). ∎

6. The Structure of Equilibrium Measures.

The long-time behavior of critical branching systems and the related problem of the classification of entrance laws have been extensively studied in the literature (e.g. Dawson (1977), Dynkin (1989c), Kallenberg (1977a), Liemant et al (1988)). Of particular interest is the question of the existence of non-trivial equilibrium measures. In this section we consider spatially homogeneous finite intensity equilibrium measures in the case of "(α,β)-systems". This involves extending the processes X_t and H_t to the case of infinite initial measures.

The symmetric stable process on $\mathbb{R}^d$ with exponent α is a Markov process $Y = (D,\mathcal{D},\mathcal{D}_{t+},\theta_t,Y_t,P^x)$ with generator $\Delta_\alpha = -(-\Delta)^{\alpha/2}$ ($\Delta_2=\Delta/2$ if $\alpha=2$) and semigroup $\{S_t\} =\{S_t^\alpha\}$. The corresponding path process, W, has inhomogeneous semigroup $\{T_{s,t}\}$. The $(Y,-\gamma\lambda^{1+\beta})$ superprocess, X_t, and historical process, H_t, which are determined by the cumulant equations

$$(6.1a)\qquad U_t\phi(x) = S_t\phi(x) - \gamma\int_0^t S_u[(U_{t-u}\phi)^{1+\beta}](x)\,du, \quad \phi\in bp\mathcal{E}$$

$$(6.1b)\qquad V_{s,t}f(y) = T_{s,t}\, f(y) - \gamma \int_s^t T_{s,u}[(V_{u,t})^{1+\beta}](y)\,du, \quad f\in bp\mathcal{D}_t,$$

will be called the (α,β) superprocess and historical process, respectively.

In order to consider the spatially homogeneous case in which $X(0) = \lambda(dx) = dx$ we must extend the definition of the basic process to a class of infinite measures. We first do this for the superprocess X_t following the construction of Iscoe (1986). Let $K_p(\mathbb{R}^d) = C_c^\infty(\mathbb{R}^d)\cup\{\phi_p\}$ where

$$\phi_p(x) = (1+ |x|^2)^{-p}, \quad x\in \mathbb{R}^d$$

Let $C_p(\mathbb{R}^d)$ (respectively $C_p^+(\mathbb{R}^d)$) be the space of continuous (respectively nonnegative, continuous) functions on $\mathbb{R}^d$ with norm

$$\|\phi\|_p = \sup_{x\in\mathbb{R}^d} |\phi(x)/\phi_p(x)|, \quad (\phi\in C_p(\mathbb{R}^d) \Leftrightarrow \|\phi\|_p < \infty).$$

Let $M_p(\mathbb{R}^d)$ denote the space of measures μ on $\mathbb{R}^d$ such that $\int\phi_p d\mu < \infty$. The measure λ belongs to $M_p(\mathbb{R}^d)$ for $p > d/2$. M_p is given the p-vague topology, that is, the smallest topology making the maps $\mu \to \int\phi d\mu$ continuous for all $\phi\in K_p(\mathbb{R}^d)$ (or equivalently, $\phi\in C_p(\mathbb{R}^d)$). We metrize $M_p(\mathbb{R}^d)-\{0\}$ so that $B \subset M_p(\mathbb{R}^d)-\{0\}$ is bounded if and only if $0 < \inf_{\mu\in B} \langle\mu,\phi_p\rangle \le \sup_{\mu\in B} \langle\mu,\phi_p\rangle < \infty$. This determines the topology on $M_{LF}(M_p(\mathbb{R}^d)-\{0\})$.

If $d/2 < p < (d+\alpha)/2$ then $K_p(\mathbb{R}^d) \subset \mathrm{dom}(\Delta_\alpha)$, where $\mathrm{dom}(\Delta_\alpha)$ is the domain of Δ_α in $C_p(\mathbb{R}^d)$. Also the operators Δ_α and S_t^α map K_p into $C_p(\mathbb{R}^d)$, and $t \longrightarrow S_t^\alpha\phi$ is a continuous curve in $C_p(\mathbb{R}^d)$ for each $\phi\in C_p(\mathbb{R}^d)$ such that $\lim_{|x|\to\infty} \phi(x)/\phi_p(x)$ exists (cf. Dawson, Gorostiza, Fleischmann (1989)). Then the argument of Iscoe (1986) yields a càdlàg $M_p(\mathbb{R}^d)$-valued extension of the superprocess X_t with laws $\{Q_\mu : \mu\in M_p(\mathbb{R}^d)\}$.

In this section we study the long-time behavior of the (α,β) superprocess X_t and historical process H_t (in $\mathbb{R}^d$) when the initial measure is $X(0) = \lambda(dx)$. Let $p\in (d/2,(d+\alpha)/2)$. Since Lebesgue measure is invariant for the symmetric stable process, it follows from (3.15) that $Q_\lambda(<X_t,\phi>) = \int \phi d\lambda < \infty$ for $\phi \in C_p(\mathbb{R}^d)$ and $t \geq 0$. This implies that the $M_p(\mathbb{R}^d)$-valued random measures X_t are tight. Integrating both sides of (6.1a) with respect to Lebesgue measure it follows that $\int U_t\phi(x)dx$ is monotone decreasing in t if $\phi \in C_p^+(\mathbb{R}^d)$. Thus the Laplace transform of $<X(t),\phi>$, is monotone in t. Together with the tightness noted above it follows that $X(t)$ converges weakly in $M_p(\mathbb{R}^d)$ to $X(\infty)$ as $t \longrightarrow \infty$ for some $M_p(\mathbb{R}^d)$-valued random measure X_∞. It can be shown that X_∞ is either the null measure or an equilibrium infinitely divisible random measure with $E(X_\infty) = \lambda$ (cf. Dawson (1977), Dynkin (1989c)). In the latter case X is said to be persistent. We first establish the persistence when $d > \alpha/\beta$ and obtain a representation for the Palm measures associated to X_∞ in this case.

Proposition 6.1. Let X_t denote the superprocess with initial measure $X_0(dx) = dx$ and let $p \in (d/2,(d+\alpha)/2)$.

(a) The Palm measure (on R^d) associated with the canonical measure R_t of X_t is given by

$$(6.2) \qquad \int e^{-<\mu,\phi>}(R_t)_x(d\mu) = P^x\left(e^{-\gamma(1+\beta)\int_0^t (U_s\phi(w_s))^\beta ds} \right), \quad \phi \geq 0.$$

(b) If $d > \alpha/\beta$ X is persistent and X_t converges weakly to X_∞ in $M_p(\mathbb{R}^d)$. Moreover the Palm distributions $(R_\infty)_x$ on $M_p(\mathbb{R}^d)$ associated with the canonical measure R_∞ of X_∞ are given by the Laplace functional

$$(6.3) \qquad \int e^{-<\mu,\phi>}(R_\infty)_x(d\mu) = P^x\left(e^{-\gamma(1+\beta)\int_0^\infty (U_u\phi(W_u))^\beta ds} \right),$$

$\phi \in C_p^+(\mathbb{R}^d)$,

where P^x is the law of the α-symmetric stable process.

<u>Proof.</u> (a) Setting $\phi(w) = f(w(t))$ $(f \geq 0)$ in (4.1.11) we obtain for $P_{0,\lambda}$-a.a. w (choose $m \in M_F(\mathbb{R}^d)$ equivalent to λ to apply (4.1.11))

(6.4)

$$\int e^{-\langle\mu,\phi\rangle}(R_{0,t})_{w^t}(d\mu) = \left(e^{-\gamma(1+\beta)\int_0^t (V_{r,t}\phi(w^r))^\beta dr} \right)$$

$$= \left(e^{-\gamma(1+\beta)\int_0^t (U_{t-r}f(w(r)))^\beta dr} \right) \quad \text{(by (2.2.9))}.$$

If $\psi(w) = g(w(t))$ $(g \geq 0)$, then

$$\text{(6.5)} \qquad \int \psi(w^t)\left(\int e^{-\langle\mu,\phi\rangle}(R_{0,t})_{w^t}(d\mu)\right)P_{0,\lambda}(dw)$$

$$= \int \langle\mu,\psi\rangle e^{-\langle\mu,\phi\rangle}R_{0,t}(d\mu) = \int \langle\nu,g\rangle e^{-\langle\nu,f\rangle}R_t(d\nu), \text{ by (4.1.7)}$$

where $R_{0,t}(d\mu) = \int R_{0,t}(x,d\mu)\lambda(dx)$ (identifying D^0 with $\mathbb{R}^d$), $P_{0,\lambda}(dw) = \int P^x(dw)\lambda(dx)$, and $R_t = \bar{\Pi}_t R_{0,t}$. But then

$$\text{(6.6)} \qquad \int \psi(w^t)\left(\int e^{-\langle\mu,\phi\rangle}(R_{0,t})_{w^t}(d\mu)\right)P_{0,\lambda}(dw)$$

$$= \int g(w(t)) \int \left(e^{-\gamma(1+\beta)\int_0^t (U_{t-r}f(w(r)))^\beta dr} \right) P_{0,\lambda}(dw)$$

by (6.4)

$$= \int g(x)\left\{\int\left(e^{-\gamma(1+\beta)\int_0^t (U_{t-r}f(w(t-r)))^\beta dr}\right)P^x(dw)\right\} dx$$

where we have used the reversibility of the α-symmetric stable process. Then (6.5) and (6.6) together with the definition of $(R_t)_x$ yields (a).

(b) We next verify that the Palm measures $(R_t)_x$ converge weakly in $M_p(\mathbb{R}^d)$ as $t\to\infty$ and that the limiting measure $(R_\infty)_x$ has Laplace functional given by (6.3). Since the Laplace functional of $(R_t)_x$ is monotone in t by (6.2), to prove convergence in $M_p(\mathbb{R}^d)$ it suffices to verify tightness. Since by (6.2) $(\phi\geq 0)$

$$\int e^{-\langle \mu,\theta\phi\rangle}(R_t)_x(d\mu) = P^x\left(e^{-\gamma(1+\beta)\int_0^t (U_s\theta\phi(w_s))^\beta ds}\right),$$

it follows that

$$(R_t)_x(\langle \phi,\mu\rangle \geq \theta^{-1}) \leq \frac{e}{e-1}\left[1 - P^x\left(e^{-(1+\beta)\int_0^t (U_s\theta\phi(w_s))^\beta ds}\right)\right].$$

But if $\phi\in C_p^+(\mathbb{R}^d)$, then using the integrability of ϕ (with respect to λ) and the unimodality and scaling properties of the α-symmetric stable process in $\mathbb{R}^d$

$$\int_0^\infty (U_s\theta\phi(w_s))^\beta ds \leq \int_0^\infty (S_s^\alpha\theta\phi(w_s))^\beta ds$$

$$\leq \text{const}\ \|\phi\|_p^\beta\ \theta^\beta\int_0^\infty (1\vee t)^{-\beta d/\alpha}dt$$

and hence is bounded for $d > \alpha/\beta$. Hence

$$(6.7)\qquad (R_t)_x(\langle \phi,\mu\rangle \geq \theta^{-1}) \leq \text{const}\ \|\phi\|_p^\beta\ \theta^\beta \quad \text{(uniform in t)}$$

and the tightness is established. (6.3) is immediate from (6.2).

It remains to show that, as our notation suggests, $(R_\infty)_x$ are the Palm distributions of the canonical measure of X_∞ and that the mean measure of X_∞ coincides with the mean measure of X_0 (i.e. λ). The weak convergence of X_t to X_∞ shows that $\forall\phi\in C_p^+(\mathbb{R}^d)$

$$\lim_{t\to\infty}\int(1-e^{-\langle\mu,\phi\rangle})R_t(d\mu) = \int(1-e^{-\langle\mu,\phi\rangle})R_\infty(d\mu).$$

From this it is easy to see that $R_t \longrightarrow R_\infty$ vaguely in $M_{LF}(M_p(\mathbb{R}^d)-\{0\})$ (cf.Liemant et al (1989, 1.6.3 and references therein)). For ϕ as above we claim that

$$\int\langle\mu,\phi\rangle R_t(d\mu) \longrightarrow \int\langle\mu,\phi\rangle R_\infty(d\mu) \quad \text{as} \quad t\to\infty.$$

(Note that $\int\langle\mu,\phi\rangle R_t(d\mu) = E(\langle X_t,\phi\rangle) = \int\phi(x)dx$ so this convergence implies the mean measure of X_∞ is Lebesgue measure.) In view of the above vague convergence, this follows from the uniform integrability

(6.8a)
$$\lim_{K\to\infty}\sup_t\int\langle\mu,\phi\rangle\mathbf{1}(\langle\mu,\phi\rangle > K)R_t(d\mu)$$
$$= \lim_{K\to\infty}\sup_t\int\phi(x)(R_t)_x(\langle\mu,\phi\rangle > K)dx \quad (\text{by } (4.1.7))$$
$$= 0 \quad (\text{by } (6.7)),$$

and for $t_0>0$

(6.8b)
$$\lim_{\varepsilon\downarrow 0}\sup_{t\ge t_0}\int\langle\mu,\phi\rangle\mathbf{1}(\langle\mu,\phi\rangle \le \varepsilon)R_t(d\mu)$$
$$\le \lim_{\varepsilon\downarrow 0}\int\phi(x)(R_{t_0})_x(\langle\mu,\phi\rangle \le \varepsilon)dx = 0$$

by the monotonicity of the Palm measures implied by Proposition 4.1.5. If $\psi\in C_b(M_p(\mathbb{R}^d))$ and $\phi\in C_p^+(\mathbb{R}^d)$ then the vague convergence of R_t to R_∞ and the above convergence of their intensities easily implies (this is the generalized Palm-Khinchin theorem (Liemant et al (1988, 1.7.9 and references therein) but the proof is elementary)

$$\int\langle\mu,\phi\rangle\psi(\mu)R_\infty(d\mu) = \lim_{t\to\infty}\int\langle\mu,\phi\rangle\psi(\mu)R_t(d\mu)$$
$$= \lim_{t\to\infty}\int\langle(R_t)_x,\psi\rangle\phi(x)\,dx$$
$$= \int\langle(R_\infty)_x,\psi\rangle\phi(x)\,dx,$$

the last by dominated convergence and the weak convergence of the Palm measures. This shows $(R_\infty)_x$ are the Palm measures of R_∞ and completes the proof. ∎

Remark. The persistence of the (2,1)-branching system in $d \geq 3$ was proved in Dawson (1977) by an analytical argument. Kallenberg (1977a) developed the backward tree method to obtain a criterion for persistence of discrete time branching systems. Using p.d.e. methods Dawson and Fleischmann (1985) proved that for $\alpha=2$, the system is persistent if and only if $d > 2/\beta$. Based on results of Gorostiza and Wakolbinger (1989) involving a modification of Kallenberg's method for particle systems in continuous time, Gorostiza, Roelly-Coppoletta and Wakolbinger (1989) showed that the (α,β)-system is persistent if and only if $d > \alpha/\beta$. The above proof is included to illustrate that such results can be obtained directly from the expression for the Palm distribution. A characterization of finite intensity entrance laws was obtained by Dynkin (1989c) for continuous branching systems and in Liemant, Matthes and Wakolbinger (1988) for discrete time branching systems.

We now proceed to discuss the equilibrium in the context of the historical process. Following Dynkin (1989c) it is natural to interpret the equilibrium random measure for the superprocess as an entrance law at $-\infty$ and to observe the process at time 0. In order to formalize this we must extend the process H_t to the time interval $(-\infty,\infty)$ and include infinite measures. The main result on the structure of the equilibrium measure is given in Theorem 6.3.

Let $D^{\pm} = D^{\pm}(\mathbb{R}^d)$ be the space of càdlàg $\mathbb{R}^d$-valued paths on $(-\infty,\infty)$ with the Skorohod J_1-topology (cf. Lindvall (1973)) and let $D^{\pm,t} = \{y\in D^{\pm}: y=y^t\equiv y(t\wedge .)\}$ $(t\in\mathbb{R})$. We also use $Y_u(y) = y_u$ and θ_u $(u\in\mathbb{R})$ to denote coordinate variables and shift operators on $D^{\pm}$ (as well as D), and if $y\in D^{\pm}$, $s\in\mathbb{R}$ and $w\in D$ define $y/s/w \in D^{\pm}$ by

$$(y/s/w)(u) = \begin{cases} y(u) & \text{if } u<s \\ w(u-s) & \text{if } u\geq s \end{cases}.$$

If $I\subset\mathbb{R}$ let $\mathcal{D}^{\pm}(I) = \sigma(Y_u:u\in I) \subset \mathcal{D}^{\pm}\equiv\mathcal{D}^{\pm}(\mathbb{R})$. If $(s,y) \in \hat{E}^{\pm}=\{(s,y)\in\mathbb{R}\times D^{\pm}:y=y^s\}$, define a probability $\tilde{P}_{s,y}$ on $(D^{\pm},\mathcal{D}^{\pm})$ by

$$\tilde{P}_{s,y}(\phi) = P^{y(s)}(\phi(y/s/Y)) \quad \phi\in b\mathcal{D}^{\pm}.$$

If $W_t^{\pm}(y) = y^t\in D^{\pm,t}$ and we extend (H) in Section 2 and the definition of IBSMP to processes indexed by $\mathbb{R}$ in the obvious way, then it is easy to see that $\{\tilde{P}_{s,y}:(s,y)\in\hat{E}^{\pm}\}$ satisfy (H) and $W^{\pm} = (D^{\pm},\mathcal{D}^{\pm},\mathcal{D}^{\pm}[s,t+],W_t^{\pm},\tilde{P}_{s,y})$ is an IBSMP with càdlàg paths in $D^{\pm,t}\subset D^{\pm}$ and inhomogeneous semigroup

$$T_{s,t}f(y) = P^{y(s)}(f(y/s/Y^{t-s})) \quad -\infty<s<t<\infty,\ f\in b\mathcal{D}^{\pm}, y\in D^{\pm}.$$

Let $\tilde{H}_t(\omega) = \omega(t)$ denote the coordinate variables on

$$(\tilde{\Omega},\tilde{\mathcal{G}}) = (D^{\pm}(M_F(D^{\pm})),\mathcal{B}(D^{\pm}(M_F(D^{\pm})))),\quad \tilde{\mathcal{G}}[s,t] = \sigma(\tilde{H}_u:s\leq u\leq t),$$

and

$$M_F(D^{\pm})^t = \{m\in M_F(D^{\pm}):m((D^{\pm,t})^c)=0\},\ t\in\mathbb{R}.$$

Now repeat the construction of the historical process in Section 2 with $W^{\pm}$ in place of W to construct probabilities $\{\tilde{Q}_{s,m}:s\in\mathbb{R},m\in M_F(D^{\pm})^s\}$ such that $\tilde{H} = (\tilde{\Omega},\tilde{\mathcal{G}},\tilde{\mathcal{G}}[s,t+],\tilde{H}_t,\tilde{Q}_{s,m})$ is an IBSMP with càdlàg paths in $M_F(D^{\pm})^t \subset M_F(D^{\pm})$. If $V_{s,t}f(y)$ $(-\infty<s\leq t<\infty,\ f\in bp\mathcal{D}^{\pm}, y\in D^{\pm})$ is the unique solution of (6.1b) (where

$T_{s,t}f$ has been extended as above) then the laws $\{\tilde{Q}_{s,m}\}$ are characterized by

$$(6.9)\quad \tilde{Q}_{s,m}(\exp\{-<\tilde{H}_t,f>\}) = \exp\{-<m,V_{s,t}f>\},\ -\infty<s\le t<\infty,\ m\in M_F(D^{\pm})^s,\ f\in bp\mathcal{D}^{\pm}.$$

Although our setting in this section is that of the (α,β) historical process $H = (\Omega,\mathcal{G},\mathcal{G}[s,t+],H_t,Q_{s,m})$, the above construction is valid under the hypotheses (M_1) and (M_2) of Section 2.2 (where (6.1b) is replaced by (2.1.16)). An easy calculation using uniqueness in (2.1.16) shows that if $\Pi_{[0,\infty)}:D^{\pm}\longrightarrow D$ is the natural projection then for some $s\ge 0$ and $m\in M_F(D^{\pm})^s$,

$$Q_{s,\bar{\Pi}_{[0,\infty)}(m)}(H_{\cdot}\in A) = \tilde{Q}_{s,m}(\bar{\Pi}_{[0,\infty)}(\tilde{H}_{\cdot})\in A),\ \forall\ A\in\tilde{\mathcal{G}}[s,\infty)$$

(note that $T_{s,t}(f\circ\Pi_{[0,\infty)})(y) = T_{s,t}f(\Pi_{[0,\infty)}(y))$ for $f\in b\mathcal{D}$, $y\in D^{\pm}$).

To consider infinite initial measures it will be convenient to introduce $\bar{D}^{\pm} = D((-\infty,\infty],\mathbb{R}^d)$ with the J_1-topology and Borel sets $\bar{\mathcal{D}}^{\pm}$, and consider $\tilde{H}$ as an IBSMP with càdlàg paths in $M_F(D^{\pm})^t \equiv M_F(\bar{D}^{\pm})^t \subset M_F(\bar{D}^{\pm})$ (i.e. $\tilde{Q}_{s,m}$ is also a probability on $(D^{\pm}(M_F(\bar{D}^{\pm})),\tilde{\mathcal{G}}[s,\infty))$ for each $m\in M_F(\bar{D}^{\pm})^s$). Note that the semigroups $T_{s,t}f$ and $V_{s,t}f$ are also defined as semigroups on $bp\bar{\mathcal{D}}^{\pm}$. Fix $p\in(d/2,(d+\alpha)/2)$ and define $\phi_{p,\infty}:\bar{D}^{\pm}\longrightarrow[0,1]$ by $\phi_{p,\infty}(y) = \phi_p(y(\infty))$. Let

$$C_p^+ = \{\phi\in C(\bar{D}^{\pm}):0\le\phi(y)\le c\phi_{p,\infty}(y) \text{ for some } c\ge 0\},$$

$$\|\phi\|_p = \sup\{\phi(y)/\phi_{p,\infty}(y):y\in\bar{D}^{\pm}\},\ \phi\in C_p^+,$$

$$M_p = M_p(\bar{D}^{\pm}) = \{m\in M(\bar{D}^{\pm}):<m,\phi_{p,\infty}> < \infty\},$$

and

$$M_p^t = M_p(\bar{D}^{\pm})^t = \{m\in M_p:m((\bar{D}^{\pm,t})^c) = 0\}.$$

We give M_p the weak topology induced by C_p^+ and $M_p^t \subset M_p$ inherits the subspace topology. M_p is then a Polish space.

Let $s \in \mathbb{R}$ and $m \in M_p^s$. Define $m_n \in M_F(\bar{D}^{\pm})^s$ by

$$m_n(A) = m(A \cap \{y \in \bar{D}^{\pm} : |y(s)| \le n\}).$$

Clearly $m_n \longrightarrow m$ in M_p^s. As in Perkins (1988, Section 7) one may define a sequence of càdlàg $M_F(\bar{D}^{\pm})$-valued processes $\{\tilde{H}_{\cdot}^{(n)}\}$ on a common $(\Omega, \mathcal{F}, P)$ such that

$$P(\tilde{H}_t^{(n)} \in \cdot) = \tilde{Q}_{s,m_n} \quad \text{on } \tilde{\mathcal{G}}[s,\infty)$$

$$\tilde{H}_t^{(m)}(\cdot) = \tilde{H}_t^{(n)}(\cdot \cap \{y : |y(s)| \le m\}) \quad \text{for } n \ge m \text{ and } t \ge s \text{ a.s.}$$

By monotonicity we may define a $M(\bar{D}^{\pm})$-valued process by

$$\tilde{H}_t^{(\infty)}(A) = \lim_{n \to \infty} \tilde{H}_t^{(n)}(A), \; A \in \bar{\mathcal{D}}^{\pm}, \; t \ge s.$$

We also let $\tilde{H}_t(\omega) = \omega(t)$ denote the coordinate variables on $(\tilde{\Omega}_p, \tilde{\mathcal{G}}_p) = (D^{\pm}(M_p), \mathcal{B}(D^{\pm}(M_p)))$ and let $\tilde{\mathcal{G}}_p[s,t] = \sigma(\tilde{H}_u : s \le u \le t) \subset \tilde{\mathcal{G}}_p$.

<u>Proposition 6.2.</u> $\{\tilde{H}_t^{(\infty)} : t \ge s\}$ is a càdlàg M_p-valued process such that $\tilde{H}_t^{(\infty)} \in M_p^t$ for all $t \ge s$ a.s. The law of $\tilde{H}^{(\infty)}$, $\tilde{Q}_{s,m}^p$ on $\tilde{\mathcal{G}}_p[s,\infty)$, satisfies

(6.10a) $\tilde{Q}_{s,m}^p(\exp(-\langle \tilde{H}_t, f \rangle)) = \exp\{-\langle m, V_{s,t} f \rangle\}, \; s \le t, \; f \in C_p^+,$

(6.10b) $\tilde{Q}_{s,m}^p(\bar{\theta}_t(\tilde{H}_t) \in A) = \tilde{Q}_{s-t,\bar{\theta}_t(m)}^p(\tilde{H}_0 \in A) \quad \forall A \in \mathcal{B}(M_p), \; s \le t,$

and

(6.10c) $V_{s,t} : C_p^+ \longrightarrow C_p^+;$

if $\phi \in C_p^+$, $\sup_{0 \le t-s \le T} |V_{s,t}\phi(y)| \le c'(T) \|\phi\|_p \phi_{p,\infty}(y);$

if $\phi \in C_b(\bar{D}^{\pm})$, $s \longrightarrow V_{s,t}\phi$ $(s \le t)$ is a continuous curve in $C_b(\bar{D}^{\pm})$.

<u>Proof.</u> It follows easily from our construction of $\tilde{Q}_{s,m_n}$ and Fitzsimmons (1989, Proposition 2.1) that

$$\langle \tilde{H}_t^{(n)}, \phi_{p,\infty} \rangle - \langle \tilde{H}_s^{(n)}, \phi_{p,\infty} \rangle - \int_s^t \langle \tilde{H}_u^{(n)}, \Delta_{\alpha,u} \phi_{p,\infty} \rangle du$$

is a càdlàg martingale in $t \geq s$, where $\Delta_{\alpha,u}\phi_{p,\infty}(y^u) = \Delta_\alpha \phi_p(y(u)) \leq c\phi_{p,\infty}(y^u)$ (cf. Dawson and Gorostiza (1990, Lemma 2.8 and remark)). It follows from (6.1b) that for $t \geq s$,

$$0 \leq \theta T_{s,t}(\phi_{p,\infty}) - V_{s,t}(\theta\phi_{p,\infty}) \leq \gamma\theta^{1+\beta}\int_s^t T_{s,u}((T_{u,t}\phi_{p,\infty})^{1+\beta})du.$$

Then using Markov's inequality (with the increasing function $x \to e^{-x}+x-1$) we obtain for $s \leq t \leq T$ and $0<\theta\leq 1$,

$$P(\langle \tilde{H}_t^{(n)}, \phi_{p,\infty}\rangle \geq \theta^{-1})$$

$$\leq e\left[\exp\{-\langle m_n, V_{s,t}(\theta\phi_{p,\infty})\rangle\} + \theta\langle m_n, T_{s,t}\phi_{p,\infty}\rangle - 1\right] \qquad \text{(by (6.9))}$$

$$\leq c(T,\gamma)\theta^{1+\beta}[1+\langle m,\phi_{p,\infty}\rangle^2]$$

(use $e^{-x} \leq 1-x+x^2/2$ and $S_t\phi_p \leq c'(T)\phi_p$ for $t \leq T$ (cf. Cor. 2.4 of Iscoe (1986))). Therefore if $0<\delta<\beta$ and $T>s$, then

$$\sup_n \sup_{s\leq t\leq T} P(\langle \tilde{H}_t^{(n)}, \phi_{p,\infty}\rangle^{1+\delta}) < \infty$$

and hence from the corresponding result for $\tilde{H}_t^{(n)}$,

$$\langle \tilde{H}_t^{(\infty)}, \phi_{p,\infty}\rangle - \langle \tilde{H}_s^{(\infty)}, \phi_{p,\infty}\rangle - \int_s^t \langle \tilde{H}_u^{(\infty)}, \Delta_{\alpha,u}\phi_{p,\infty}\rangle du, \quad t \geq s$$

is a P-martingale. A standard argument (see the derivation of (7.6) in Perkins (1988) but use the strong $L^{1+\delta}$ inequality in place of the square function argument given there) shows that

$$\lim_{n\to\infty} P(\sup_{s\leq t\leq T} \langle \tilde{H}_t^{(\infty)} - H_t^{(n)}, \phi_{p,\infty}\rangle^{1+\delta}) = 0.$$

By monotonicity in n it follows that

$$\sup_{s\leq t\leq T} |\langle \tilde{H}_t^{(\infty)}, \phi\rangle - \langle \tilde{H}_t^{(n)}, \phi\rangle| \longrightarrow 0 \quad \forall\ \phi \in C_p^+ \text{ and } T>s \quad \text{P-a.s.}$$

Therefore $\tilde{H}_t^{(\infty)}$ $(t \geq s)$ is a càdlàg M_p-valued process and $\tilde{H}_t^{(\infty)} \in M_p^t$ for all $t \geq s$ since $\tilde{H}_t^{(n)}((\bar{D}^{\pm,t})^c) = 0$ for all $t \geq s$ and $n \in \mathbb{N}$ a.s. (6.10a) follows by setting $m=m_n$ in (6.9) and letting $n \longrightarrow \infty$.

(6.10b) is obvious from the probabilistic description since it just amounts to shifting all the paths by t. To prove it analytically note that $T_{s-t,0}\phi(\theta_t y) = T_{s,t}(\phi\circ\theta_t)(y)$ for $s \leq t$,

$y\in D$, $\phi\in bp\mathcal{D}^{\pm}$, by a trivial calculation. The uniqueness in (6.1b) now implies

(6.11) $V_{s-t,0}\phi(\theta_t y) = V_{s,t}(\phi\circ\theta_t)(y)$ $s\le t$, $y\in D$, $\phi\in bp\mathcal{D}^{\pm}$.

(6.10b) is now immediate from (6.10a) and (6.11).

Using $S_t\phi_p \le c'(T)\phi_p$ $(t\le T)$ again it is easy to see that for $\phi\in C_p^+$

$$V_{s,t}\phi(y) \le T_{s,t}\phi(y) \le c'(T)\|\phi\|_p\phi_{p,\infty}(y) \quad \text{for} \quad t-s \le T.$$

The other properties in (6.10c) are immediate from the classical construction of $V_{s,t}\phi$ via fixed point theorems. ∎

By using the Markov property of $\{\tilde{Q}_{s,m}: m\in M_F(\bar{D}^{\pm})^s, s\in\mathbb{R}\}$ and the continuity of the right side of (6.10a) in $(s,m)\in \hat{M}_p = \{(s,m)\in \mathbb{R}\times M_p : m\in M_p^s\}$ (use (6.10c)) it is easy to use the above construction of $\tilde{Q}_{s,m}$ as a limit of $\{\tilde{Q}_{s,m_n}\}$ to see that $\tilde{H} = (\tilde{\Omega}_p, \tilde{\mathcal{G}}_p, \tilde{\mathcal{G}}_p[s,t+], \tilde{H}_t, \tilde{Q}^p_{s,m})$ is an IBSMP with càdlàg paths in $M_p^t \subset M_p$. Hence the laws $\{\tilde{Q}^p_{s,m}: (s,m)\in\hat{M}_p\}$ are uniquely determined by (6.10a).

For $-\infty<s<\infty$ and $A\in \mathcal{D}^{\pm}[s,\infty)$ define

$$P_s(A) = \int P^x(Y((\cdot-s)^+)\in A)\lambda(dx).$$

Since λ is invariant for the stable process Y, the family $\{P_s: s\in\mathbb{R}\}$ is consistent and hence defines a unique measure $P_{-\infty}$ on $(D^{\pm},\mathcal{D}^{\pm})$ by

$$P_{-\infty}(A) = P_s(A) \quad \forall\, A\in \mathcal{D}^{\pm}([s,\infty)),\ s\in \mathbb{R}.$$

Define $\tilde{\lambda}\in \bigcap_{t\in\mathbb{R}} M_p^t$ by

$$\tilde{\lambda}(\{y: y(0)\in B\}) = \lambda(B), \quad \tilde{\lambda}(\{y: y(t)=y(0)\ \forall t\in (-\infty,\infty]\}^c) = 0.$$

A Borel set $C\subset \bar{D}^{\pm}$ is a clan history if there is a y_0 in $\bar{D}^{\pm}$ such that $\forall y$ in C there is a $\tau(y)\in \mathbb{R}$ such that $y(s)=y_0(s)$ for all $s<\tau(y)$. A measure m in M_p is a clan measure if it is supported by a clan history. The set of clan measures in M_p^t, denoted by $M_p^{cl,t}$, is a measurable subset (see Lemma A.6).

<u>Theorem 6.3.</u> Let $d > \alpha/\beta$.

(a) For each $t \in \mathbb{R}$ $\tilde{Q}^p_{s,\tilde{\lambda}}(\tilde{H}_t \in .)$ converges weakly in M_p^t as $s \downarrow -\infty$ to an infinitely divisible random measure $\tilde{H}_t^{-\infty}$ ($\in M_p^t$) with intensity

$$E(\tilde{H}_t^{-\infty}(A)) = P_{-\infty}(Y^t \in A) \quad \forall A \in \bar{\mathcal{D}}^{\pm}$$

and Laplace functional

$$(6.12a) \quad E(\exp\{-<\tilde{H}_t^{-\infty},\phi>\}) = \exp\{-\bar{V}_t(\phi)\} \quad \phi \in C_p^+ \cap b\mathcal{D}^{\pm}[s_0,\infty] \text{ for some } s_0 > -\infty,$$

where

$$(6.12b) \quad \bar{V}_t(\phi) = \lim_{s \to -\infty} \int (V_{s,t}\phi)(y)\tilde{\lambda}(dy), \quad \phi \text{ as in (6.12a).}$$

The canonical measure $R_{-\infty,t}$ of $\tilde{H}_t^{-\infty}$ on M_p^t therefore satisfies

$$(6.12c) \quad \int (1-e^{-<m,\phi>})R_{-\infty,t}(dm) = \bar{V}_t\phi, \quad \phi \text{ as in (6.12a).}$$

$\bar{\Pi}_t(\tilde{H}_t^{-\infty})$ is equal in law to X_∞, the weak limit of $\{X_t\}$ in Proposition 6.1.

(b) The Palm distributions associated to $R_{-\infty,t}$ have Laplace functionals

$$(6.13) \quad \int e^{-<m,\phi>}(R_{-\infty,t})_y(dm) = \exp\{-\gamma(1+\beta)\int_{-\infty}^t (V_{r,t}\phi(y^r))^\beta dr\}$$

$$\phi \text{ as in (6.12a).}$$

(c) For $P_{-\infty}$ a.a. y

$$(R_{-\infty,t})_{y^t}((M_p^{cl,t})^c) = 0$$

and

$$R_{-\infty,t}((M_p^{cl,t})^c) = 0.$$

<u>Proof.</u> (a),(b) If ϕ is as in (6.12a) and $s \le s_0$ it is easy to see that

$$\tilde{Q}^p_{s,\tilde{\lambda}}(<\tilde{H}_t,\phi>) = P_s(\phi(Y^t)) = P_{-\infty}(\phi(Y^t)) < \infty.$$

A standard (or nonstandard) argument now shows that $\{\tilde{Q}_{s,\tilde{\lambda}}(\tilde{H}_t\in\cdot):s\le s_0\}$ are tight measures on M_p^t. As in the proof of Proposition 6.1 one can use (6.1b) to see that the Laplace transforms

$$\tilde{Q}^p_{s,\tilde{\lambda}}(\exp\{-\langle\tilde{H}_t,\phi\rangle\}) = \exp\{-\int V_{s,t}\phi\, d\tilde{\lambda}\}$$

are decreasing in $s\le s_0$. The weak convergence of $\tilde{Q}^p_{s,\tilde{\lambda}}(\tilde{H}_t\in\cdot)$ in M_p^t as $s\downarrow-\infty$ and (6.12a,b) follow. As in the proof of Proposition 6.1, to complete the proof it suffices to show weak convergence in M_p^t of the Palm distributions $(R_{s,t})_y$ of the canonical measure $R_{s,t}$ of $\tilde{Q}^p_{s,\tilde{\lambda}}(\tilde{H}_t\in\cdot)$ as $s\downarrow-\infty$, and to establish the analogue of the uniform integrability (6.8). By Proposition 4.1.5 random measures $\{\xi_{s,y}:s\le t\}$ with laws $(R_{s,t})_y$ and Laplace functionals

$$\int e^{-\langle\phi,m\rangle}(R_{s,t})_y(dm) = \exp\left\{-\gamma(1+\beta)\int_s^t (V_{r,t}\phi(y^r))^{\beta}dr\right\}$$

can be realized on a common probability space in such a way that $\xi_{s,y}\uparrow$ as $s\downarrow-\infty$. This monotonicity allows us to derive the tightness in M_p^t as in the derivation of (6.7) by noting that if $\phi\in C_p^+$, then

$$V_{s,t}\phi(y) \le U_{t-s}(c\phi_p)(y(s)) \quad \text{for some } c>0.$$

The uniform integrability follows in a similar manner.

If $\phi = \psi\circ\Pi_t$ where $\psi\in C_p^+(\mathbb{R}^d)$ then (6.12b) implies $\bar{V}_t(\phi)$ $= \lim_{u\to\infty}\int U_u\phi(x)dx$ and so by (6.12a) $\bar{\Pi}_t(\tilde{H}_t^{-\infty})$ and X_∞ have identical Laplace functionals.

(c) The result for the Palm measures $(R_{-\infty,t})_{y^t}$ follows from (6.13) and the obvious extension of the representation in Proposition 4.1.5. The result for $R_{-\infty,t}$ then follows from the analogue of (4.1.7).∎

The process $\tilde{H}_t$ under $\tilde{Q}^p_{s,m}$ induces a clan-valued process $\xi_t \in M_p^{cl,t}$ which represents the evolution of one of the unrelated clans which constitutes $\tilde{H}_t$. More precisely, $\{\xi_t : t \geq s\}$ is defined to be the process $\{H_t : t \geq s\}$ when the initial measure $H_s \in M_p^{cl,s}$. Therefore it is an M_p-valued IBSMP whose transition function is given by (use (3.4) and (3.7))

$$(6.14a) \quad \hat{Q}_{s,\xi_s}(e^{-<\xi_t,\phi>}) = \exp\{-<\xi_s, V_{s,t}\phi>\} \quad \phi \in C_p^+.$$

In Lemma A.9(a) we verify that for each $\xi_s \in M_p^{cl,s}$,

$$\hat{Q}_{s,\xi_s}(\xi_t \in M_p^{cl,t} \ \forall t \geq s) = 1$$

and therefore it is an M_p^{cl}-valued IBSMP.

Next we introduce a point process of clan measures $(\Xi_t, t \geq 0)$ to describe the ensemble of clan measures which constitute $H_t^{-\infty}$. We can (and shall) introduce a metric on $M_p - \{0\}$ such that B is bounded in $M_p - \{0\}$ if and only if

$$0 < \inf_{\mu \in B} <\mu, \phi_{p,\infty}> \leq \sup_{\mu \in B} <\mu, \phi_{p,\infty}> < \infty.$$

Then $R_{-\infty,t} \in M_{LF}(M_p^{cl,t} - \{0\})$ since

$$\int <\mu, \phi_{p,\infty}> R_{-\infty,t}(d\mu) = E(<\tilde{H}_t^{-\infty}, \phi_{p,\infty}>) < \infty \text{ (by Theorem 6.3(a))}.$$

Hence we may introduce a Poisson random measure Ξ_0 in $M_p^{cl,0}$ with intensity $R_{-\infty,0}$. $(\Xi_t, t \geq 0)$ is defined to be the $M_{LF}(M_p^{cl,t})$-valued point process obtained by letting the initial points move as independent copies of the clan-valued process ξ_t. Hence Ξ_t is the inhomogeneous $M_{LF}(M_p^{cl})$-valued Markov process with initial law Ξ_0 and whose transition function is given by

$$(6.14b) \quad \bar{Q}_{s,\Xi_s}\left(e^{-<\Xi_t,\psi>}\right) = \exp\{<\Xi_s, \log \hat{Q}_{s,\cdot}(e^{-\psi(\xi_t)})>\},$$

$$\psi \in bp\mathcal{B}(M_p).$$

Let $\bar{Q}$ denote the law of $\Xi_{\cdot}$ on the canonical space of paths.

Given a random measure Ψ on M_p^t the projection of Ψ, defined by $\mathcal{J}(\Psi) = \int \mu\Psi(d\mu)$, yields a random measure on $\bar{D}^{\pm,t}$. Also $\bar{\theta}_t : M_p^t \longrightarrow M_p^0$ is defined by $\langle \bar{\theta}_t\mu,\phi\rangle = \langle \mu,\phi\circ\theta_t\rangle$.

Finally we use $\tilde{H}_0^{-\infty}$ as a random initial measure to construct a stationary M_p-valued process (see Theorem 6.4 below). Let $\tilde{Q}^p$ be the measure on $(\tilde{\Omega}_p,\tilde{\mathcal{G}}_p[0,\infty))$ defined by

$$\tilde{Q}^p = E\left(\int_{M_p} \tilde{Q}^p_{0,m}\ \tilde{H}_0^{-\infty}(dm)\right).$$

__Theorem 6.4.__ Let $d > \alpha/\beta$.

(a) If s<t then for any $\xi \in M_p^{cl,s}$

$$\hat{Q}_{s,\xi}(\bar{\theta}_t(\xi_t)\in\cdot) = \hat{Q}_{s-t,\bar{\theta}_t\xi}(\xi_0\in\cdot) \quad \text{on} \quad \mathcal{B}(M_p^{cl,0}).$$

In particular for any $\xi \in M_p^{cl,0}$ $\{\bar{\theta}_t\xi_t : t\geq 0\}$ is a homogeneous $M_p^{cl,0}$-valued Markov process under $\hat{Q}_{0,\xi_{\cdot}}$. If ξ_0 is given the initial (infinite) law $R_{-\infty,0}$, then $\{\bar{\theta}_t\xi_t : t\geq 0\}$ is a stationary process.

(b) The $M_{LF}(M_p^{cl,0})$-valued process $\bar{\theta}_t\Xi_t$ (obtained by applying $\bar{\theta}_t$ to the points of Ξ_t) is a stationary, homogeneous Markov process.

(c) There is a $M_{LF}(M_p^{cl})$-valued process $\{\Xi_t^* : t\geq 0\}$ defined on the probability space $(\tilde{\Omega}_p,\tilde{\mathcal{G}}_p[0,\infty),\tilde{Q}^p)$ such that

$$\tilde{H}_t = \mathcal{J}(\Xi_t^*) \quad \forall t\geq 0,\ \tilde{Q}^p\text{-a.s.}$$

Further the processes $\{\Xi^*(t):t\geq 0\}$ (under $\tilde{Q}_p$) and $\{\Xi(t):t\geq 0\}$ (under $\bar{Q}$) are equal in law.

(d) Under $\tilde{Q}^p$, $\bar{\theta}_t\tilde{H}_t$ and $\Pi_0(\bar{\theta}_t\tilde{H}_t)$ $(t\geq 0)$ are stationary M_p^0-valued and $M_p(\mathbb{R}^d)$-valued homogeneous Markov processes, respectively.

__Remark.__ (d) shows that $\tilde{H}_0^{-\infty}$ and X_∞ (equal in law to $\bar{\Pi}_0(\tilde{H}_0^{-\infty})$ by Theorem 6.3(a)) are equilibrium measures.

__Proof.__ (a) If $\phi\in C_p^+$, s<t and $\xi\in M_p^{cl,s}$ then

$$\hat{Q}_{s,\xi}\left(e^{-<\bar{\theta}_t(\xi_t),\phi>}\right) = \exp\{-<\xi, V_{s,t}(\phi\circ\theta_t)>\} \quad \text{(by (6.14a))}$$

$$= \exp\{-<\bar{\theta}_t(\xi), V_{s-t,0}\phi>\} \quad \text{(by (6.11))}$$

$$= \hat{Q}_{s-t,\bar{\theta}_t(\xi)}(e^{-<\xi_0,\phi>}) \quad \text{(by (6.14a))}.$$

This gives the equivalence in (a). It follows that for $\psi \in b\mathcal{B}(M_p)$ and $\xi \in M_p^{cl,0}$,

$$\hat{Q}_{0,\xi}(\psi(\bar{\theta}_t\xi_t)\,|\,\xi_u, u\le s) = \hat{Q}_{s,\xi_s}(\psi(\bar{\theta}_t\xi_t)) = \hat{Q}_{s-t,\bar{\theta}_{t-s}(\bar{\theta}_s\xi_s)}(\psi(\xi_0))$$

which demonstrates the homogeneous Markov property.

Let $\phi \in C_p^+ \cap b\mathcal{D}^{\pm}[s_0,\infty)$ for some $s_0 > -\infty$. From (6.14a) we have

$$\hat{Q}_{0,\xi_0}\left(1 - e^{-<\xi_t,\phi\circ\theta_t>}\right) = 1 - \exp(-<\xi_0, V_{0,t}(\phi\circ\theta_t)>)$$

and hence

$$\int \hat{Q}_{0,\xi_0}\left(1 - e^{-<\bar{\theta}_t\xi_t,\phi>}\right) R_{-\infty,0}(d\xi_0)$$

$$= \int (1 - \exp\{-<\xi_0, V_{0,t}(\phi\circ\theta_t)>\}) R_{-\infty,0}(d\xi_0)$$

$$= \bar{V}_0(V_{0,t}(\phi\circ\theta_t)) \qquad \text{(by (6.12c))}$$

$$= \lim_{T\to\infty} \int (V_{-T,0}V_{0,t}(\phi\circ\theta_t))(y)\tilde{\lambda}(dy) \qquad \text{(by (6.12b))}$$

$$= \lim_{T\to\infty} \int V_{-T-t}\phi(\theta_t y)\tilde{\lambda}(dy) \qquad \text{(by (6.11))}$$

$$= \bar{V}_0\phi \qquad \text{(because } \bar{\theta}_t\tilde{\lambda} = \tilde{\lambda})$$

$$= \int (1 - e^{-<\xi,\phi>}) R_{-\infty,0}(d\xi) \quad \text{(by (6.12c))}.$$

In view of the homogeneous Markov property established above, the stationarity of $\bar{\theta}_t\xi_t$ follows.

(b) Let $\psi \in bp\mathcal{B}(M_p)$. If $0\le s<t$ then

$$Q\left(e^{-<\bar{\theta}_t \Xi_t, \psi>}\,|\,\Xi_u, u\le s\right)$$

$$= \exp\{\langle\Xi_s, \log \hat{Q}_{s,\cdot}(e^{-\psi(\bar{\theta}_t\xi_t)})\rangle\} \quad \text{(by (6.14b))}$$

$$= \exp\{\langle\Xi_s, \log \hat{Q}_{s-t,\bar{\theta}_{t-s}\bar{\theta}_s(\cdot)}(e^{-\psi(\xi_0)})\rangle\} \quad \text{(by (a))}$$

$$= \exp\{\langle\bar{\theta}_s(\Xi_s), \log \hat{Q}_{s-t,\bar{\theta}_{t-s}(\cdot)}(e^{-\psi(\xi_0)})\rangle\}.$$

This gives the homogeneous Markov property of $\bar{\theta}_s(\Xi_s)$. To prove stationarity set $s=0$ in the first line of the above and integrate out Ξ_0 to get

$$\bar{Q}(e^{-\langle\bar{\theta}_t\Xi_t,\psi\rangle}) = \exp\{-\int(1-\hat{Q}_{0,\xi}(e^{-\psi(\bar{\theta}_t\xi_t)}))R_{-\infty,0}(d\xi)\}$$

$$= \exp\{-\int(1-e^{-\psi(\xi)})R_{-\infty,0}(d\xi)\} \quad \text{(by (a))}$$

$$= \bar{Q}(e^{-\langle\Xi_0,\psi\rangle}).$$

(c) We first check that the laws of $\tilde{H}_0$ and $\mathcal{J}(\Xi_0)$ agree. By (3.3)

$$\bar{Q}(e^{-\langle\Xi_0,\phi\rangle}) = \exp\{-\int(1-e^{-\phi(\mu)})R_{-\infty,0}(d\mu)\} \quad \text{for } \phi\in bp\mathcal{B}(M_p)_b.$$

If $\psi\in C_p^+$ and

$$\phi^\varepsilon(\mu) = \langle\mu,\psi\rangle 1(\varepsilon\le\langle\mu,\phi_{p,\infty}\rangle\le\varepsilon^{-1}) \in bp\mathcal{B}(M_p)_b$$

then we set $\phi=\phi^\varepsilon$ in the above and let $\varepsilon\downarrow 0$ (use $\int\langle\mu,\phi_{p,\infty}\rangle R_{-\infty,0}(d\mu) < \infty$) to obtain

$$\bar{Q}\left(e^{-\int\langle\mu,\psi\rangle\Xi_0(d\mu)}\right) = \exp\{-\int(1-e^{-\langle\mu,\psi\rangle})R_{-\infty,0}(d\mu)\}$$

$$= \tilde{Q}^p(\exp\{-\langle\tilde{H}_0,\psi\rangle\}).$$

Since Ξ_0 is given by a countable collection of clan measures and $\tilde{H}_0$ has the same law as $\mathcal{J}(\Xi_0)$, $\tilde{H}_0$ has a decomposition of the form

$$\tilde{H}_0(\cdot) = \sum_{j=1}^{\infty} \tilde{H}_{0,j}(\cdot)$$

where $\{\tilde{H}_{0,j}\}$ are clan measures. Therefore there exist distinct random clans $\{C_j\}$ in $\mathcal{B}(\bar{D}^{\pm,0})$ (cf. proof of Lemma A.6(a)) such that

$$\tilde{H}_0(\cdot) = \sum_{j=1}^{\infty} \tilde{H}_0(\cdot \cap C_j) \qquad \tilde{Q}^p\text{-a.s.}$$

Define $\xi_t^{(j)}(A) = \tilde{H}_t(A \cap \{y : y^0 \in C_j\})$, $A \in \bar{\mathcal{D}}^{\pm}$, $t \geq 0$. By the obvious extension of Theorem 2.1.9 the processes $\{\xi_{\cdot}^{(j)}\}$ are independent and each $\xi_{\cdot}^{(j)}$ is a clan process with transition function given by (6.14a), and by Lemma A.9(a) $\xi_t^{(j)} \in M_p^{cl,t}$ $\forall t \geq 0$, $\tilde{Q}^p$-a.s. Therefore the $M_{LF}(M_p^{cl})$-valued processes

$$t \longrightarrow \Xi_t^* = \sum_{j=1}^{\infty} \delta_{\xi_t^{(j)}} \qquad \text{and} \qquad t \longrightarrow \Xi_t$$

have identical laws. (The fact that $\Xi_t^* \in M_{LF}(M_p)$ $\forall t \geq 0$, $\tilde{Q}^p$-a.s. follows since $\langle \mathcal{J}(\Xi_t^*), \phi_{p,\infty} \rangle \leq \langle \tilde{H}_t, \phi_{p,\infty} \rangle < \infty$ $\forall t \geq 0$, $\tilde{Q}^p$-a.s.)

If $\psi \in C_p^+$ and $0 \leq s < t$ then

$$\bar{Q}\left(e^{-\int \langle \mu, \psi \rangle \Xi_t(d\mu)} \,\Big|\, \Xi_u, u \leq s \right)$$

$$= \exp\{\langle \Xi_s, \log \hat{Q}_{s,\cdot}(e^{-\langle \cdot, \psi \rangle}) \rangle\} \qquad \text{(by (6.14b))}$$

$$= \exp\{-\langle \Xi_s, \langle \cdot, V_{s,t}\psi \rangle \rangle\} \qquad \text{(by (6.14a))}$$

$$= \exp\{-\langle \mathcal{J}(\Xi_s), V_{s,t}\psi \rangle\}$$

$$= \tilde{Q}^p_{s,\mathcal{J}(\Xi_s)}(e^{-\langle \tilde{H}_t, \psi \rangle}) \qquad \text{(6.10a).}$$

This shows that $\{\mathcal{J}(\Xi_t^*) : t \geq 0\}$ and $\{H_t : t \geq 0\}$ have the same finite dimensional distributions. Since $\mathcal{J}(\Xi_t^*) \leq H_t$, this implies that $H_t = \mathcal{J}(\Xi_t^*)$, $\tilde{Q}^p$-a.s. In Appendix A.9(b) we verify that $\tilde{H}_t = \mathcal{J}(\Xi_t^*)$ $\forall t \geq 0$, $\tilde{Q}^p$-a.s. which completes the proof of (c).

(d) The stationarity is immediate from (b) and (c) since $\bar{\theta}_t \tilde{H}_t$ $\overset{\mathcal{D}}{=} \mathcal{J}(\bar{\theta}_t \Xi_t)$

Let $\phi \in C_p^+$. Then

$$\tilde{Q}^{p}\left(e^{-\langle\bar{\theta}_t\tilde{H}_t,\phi\rangle}\big|\mathcal{G}^{p}_{[0,s+]}\right)$$

$$= \tilde{Q}^{p}_{s,\tilde{H}_s}\left(e^{-\langle\tilde{H}_t,\phi\circ\theta_t\rangle}\right)$$

$$= \exp\{-\langle\tilde{H}_s, V_{s,t}(\phi\circ\theta_t)\rangle\} \qquad \text{(by (6.10a))}$$

$$= \exp\{-\langle\bar{\theta}_s(\tilde{H}_s), (V_{s-t,0}\phi)\circ\theta_{t-s}\rangle\} \quad \text{(by (6.11))}.$$

This proves the homogeneous Markov property of $\bar{\theta}_t\tilde{H}_t$ and a similar argument goes through for $\Pi_0(\bar{\theta}_t\tilde{H}_t)$. (Of course for the homogeneous Markov property the initial law is irrelevant!) ∎

We next turn to the scaling properties of the canonical and Palm and measures.

Notation. $y_R(t) = y(R^{\alpha}t)/R$ and $f_R(y) = f(y_R)$, $f\in bp\mathcal{D}^{\pm}$, $y\in D^{\pm}$.

Lemma 6.5 Let $s < t$. Then the following scaling relations are satisfied:

(a)

(6.15) $$(T_{s/R^{\alpha},t/R^{\alpha}}\, f)(y_R) = T_{s,t}f_R(y).$$

(b)

(6.16) $$(R^{-\alpha/\beta}(V_{s/R^{\alpha},0}f))_R(y) = V_{s,0}\left(R^{-\alpha/\beta}\, f_R\right)(y).$$

Proof. (a) By (2.2.1) $T_{s,t}f(y) = P^{y(s)}(f(y/s/Y^{t-s}))$ and hence

$$(T_{s/R^{\alpha},t/R^{\alpha}}\, f)(y_R) = P^{y(s)/R}\left(f(y_R/(s/R^{\alpha})/Y^{(t-s)/R^{\alpha}})\right)$$

$$= P^{y(s)}\left(f(y_R/(s/R^{\alpha})/(Y^{(t-s)})_R)\right)$$

(scaling property of symmetric α-stable process)

$$= T_{s,t}f_R(y).$$

(b) From (6.1b),

$$(R^{-\alpha/\beta}V_{s/R^\alpha,0}\, f)_R(y) = (R^{-\alpha/\beta}V_{s/R^\alpha,0}\, f)(y_R)$$

$$= R^{-\alpha/\beta}(T_{s/R^\alpha,0}f)(y_R)$$

$$- \gamma\int_{s/R^\alpha}^{0} T_{s/R^\alpha,u}\Big(R^\alpha (R^{-\alpha/\beta}V_{u,0}f)^{1+\beta}\Big)(y_R)du$$

$$= R^{-\alpha/\beta}\ (T_{s,0}f_R)(y)$$

$$- \gamma\int_{s}^{0} T_{s,v}\Big((R^{-\alpha/\beta}V_{v/R^\alpha,0}f)\Big)_R^{1+\beta}(y)dv \qquad \text{by (a).}$$

Hence both sides of (6.16) satisfy (6.1b) and hence the result follows by the uniqueness of the solution. ∎

<u>Theorem 6.6</u> Let $d > \alpha/\beta$. Then the canonical measure $R_{-\infty,0}$ is α/β-self-similar, that is,

(6.17) $$R_{-\infty,0}(\{\mu:\mu_R\in A\}) = R^{d-\alpha/\beta}\, R_{-\infty,0}(A)$$

where $\mu_R(B) = R^{-\alpha/\beta}\mu(\{w:w_R\in B\})$.

<u>Proof.</u> We begin with some observations on the similarity behavior of the solutions to the cumulant equation. From Lemma 6.5 it follows that for $s<0$,

(6.18) $$(R^{-\alpha/\beta}(V_{s/R^\alpha,0}f))(y_R) = V_{s,0}\Big(R^{-\alpha/\beta}f_R\Big)(y).$$

If $f\in \mathcal{D}^{\pm}[s_0/R^\alpha,0]\cap C_p^+$, then by integration with respect to $P_{-\infty}(dy)$ we obtain for $s\le s_0$

$$\int V_{s,0}\Big(R^{-\alpha/\beta}f_R\Big)(y)P_{-\infty}(dy) = \int R^{-\alpha/\beta}(V_{s/R^\alpha,0}f)(y_R)P_{-\infty}(dy).$$

The uniqueness in (6.1b) and measurability hypothesis on f imply the integrands are measurable functions of $y(s)$ alone and therefore the above implies

$$\int V_{s,0}\left(R^{-\alpha/\beta}f_R\right)(y)\tilde{\lambda}(dy) = \int R^{d-\alpha/\beta}(V_{s/R^\alpha,0}f)(y)\tilde{\lambda}(dy).$$

Hence for such f, letting $s \to -\infty$,

$$\bar{V}_0(R^{-\alpha/\beta}f_R) = \bar{V}_0(f)R^{d-\alpha/\beta}. \tag{6.19}$$

Then by (6.12c) we conclude that $R_{-\infty,0}$ is an α/β-self-similar canonical measure that is,

$$R_{-\infty,0}(\{\mu:\mu_R\in A\}) = R^{d-\alpha/\beta}\, R_{-\infty,0}(A). \tag{6.20}$$ ∎

<u>Theorem</u> <u>6.7</u> Let $d > \alpha/\beta$ and X_0 be the random measure on $\mathbb{R}^d$ whose law is given by the Palm distribution $(R_\infty)_0$ (with Laplace functional given by (6.3)). Then X_0 is α/β-self-similar, that is, $R^{-\alpha/\beta}X_0(RA)$ is equal in law to $X_0(A)$.

<u>Proof.</u> Let $g\in bp\mathcal{B}(\mathbb{R}^d)$ and $g_R(\cdot) = R^{-\alpha/\beta}g(\cdot/R)$. From Lemma 6.5(b)

$$U_t g_R(x) = R^{-\alpha/\beta}U_{t/R^\alpha}\, g(x/R).$$

Then

$$E\left(e^{-\langle X_0,g_R\rangle}\right) = P^0\left(e^{-\gamma(1+\beta)\int_0^\infty (U_t g_R(y_t))^\beta dt}\right) \quad \text{(by (6.3))}$$

$$= P^0\left(e^{-\gamma(1+\beta)\int_0^\infty (R^{-\alpha/\beta}U_{t/R^\alpha}\, g(y_t/R))^\beta dt}\right)$$

$$= P^0\left(e^{-\gamma(1+\beta)\int_0^\infty R^{-\alpha}(U_{t/R^\alpha} g(Y_{t/R^\alpha}))^\beta dt} \right)$$

$$= P^0\left(e^{-\gamma(1+\beta)\int_0^\infty (U_s g(Y_s))^\beta ds} \right)$$

$$= E\left(e^{-\langle X_0, g\rangle} \right). \quad \blacksquare$$

<u>Definition.</u> The carrying dimension of a measure μ on $\mathbb{R}^d$ is γ if there is a Borel set $C \subset \mathbb{R}^d$ with $\mu(C^c) = 0$ and $\dim C \le \gamma$, and $\mu(B) = 0$ whenever $\dim B < \gamma$. We write $\dim(\mu) = \gamma$. (If $\mu \ne 0$ then $\dim(\mu)$ is unique if it exists.)

<u>Corollary 6.8.</u> Let $d > \alpha/\beta$. Then for R_∞-a.a. μ and $(R_\infty)_0$-a.a. μ the carrying dimension of μ is α/β.

<u>Proof.</u> This follows immediately from Theorem 3.1 of Zähle (1988a) and Theorems 6.6 and 6.7. $\blacksquare$

Recall that X_t is the (α,β) superprocess starting at $m_0 \in M_F(\mathbb{R}^d)$ under Q_{m_0}.

<u>Theorem 6.9.</u> If $t>0$ and $m_0 \in M_F(\mathbb{R}^d)$ then

$$\dim(X_t) = \alpha/\beta \quad Q_{m_0}\text{-a.s.}$$

<u>Proof.</u> Corollary 6.8 and the obvious translation invariance shows that $\dim(\mu) = \alpha/\beta$ $(R_\infty)_x$-a.a. μ for all $x \in \mathbb{R}^d$. The stochastic monotonicity implicit in Proposition 6.1, i.e. $(R_t)_x \le (R_\infty)_x$ shows that $\dim(\mu) = \alpha/\beta$ $(R_t)_x$-a.a. μ for all $t>0$ and $x \in \mathbb{R}^d$. Here $(R_t)_x$ [illegible] 6.1. It follows (e.g. by (4.1.7)) that $\dim(\mu) = \alpha/\beta$ for R_t-a.a. μ. If $R_t(x,d\mu)$ is the canonical

measure for $Q_{\delta_x}(X_t \in .)$, then, comparing Laplace functionals, we find $R_t(A) = \int_{\mathbb{R}^d} R_t(x,A)dx$. This shows that

$$\dim(\mu) = \alpha/\beta \quad \text{for } R_t(x,.)\text{-a.a. } \mu \text{ and } \lambda\text{-a.a. } x.$$

Translation invariance again allows us to conclude

$$(6.21) \qquad \dim(\mu) = \alpha/\beta \quad \text{for } R_t(x,.)\text{-a.a. } \mu \ \forall\ x \in \mathbb{R}^d,\ t>0.$$

Since X_t (under Q_{m_0}) is a Poisson cluster random measure with cluster law $R_t(x,.)/R_t(x,M_F(\mathbb{R}^d))$ and finite intensity $\Lambda(dx) = R_t(x,M_F(\mathbb{R}^d))dm_0(x)$ (see Proposition 3.3(a)) an elementary argument shows that (6.21) implies

$$\dim(X_t) = \alpha/\beta \quad Q_{m_0}\text{-a.s.} \quad \forall\ t>0,\ m_0 \in M_F(\mathbb{R}^d). \blacksquare$$

The final topic of this section is a simplified description of the equilibrium random measure X_∞ in the case $\alpha = 2$.

Proposition 6.10. Let $d > 2/\beta$, $\alpha = 2$ and $r > 0$. Then

(a) The canonical measure R_∞ of X_∞, satisfies

$$(6.22) \qquad R_\infty(\{\mu:\mu(B(0,r)) > 0\}) = c.r^{d-2/\beta} \equiv v_r$$

where $0 < c < \infty$.

(b) X_∞ restricted to $B(0,r)$ is the superposition of a Poisson number, with mean $c.r^{d-2/\beta}$, of localized clan measures. The law of a localized clan measures on $M_F(B(0,r))$ is defined by

$$(6.23) \qquad P_r^*(A) = \frac{R_\infty(A\cap\{\mu:\mu(B(0,r))>0\})}{R_\infty(\{\mu:\mu(B(0,r)>0\})}, \quad A \subset \mathcal{B}(M_F(B(0,r))),$$

and has Laplace functional

$$(6.24) \qquad \int e^{-\langle\mu,\phi\rangle} P_r^*(d\mu) = 1 - \frac{\bar{V}_0(\phi\circ\Pi_0)}{v_r}, \qquad \text{supp } \phi \subset B(0,r).$$

Proof. (a) The equivalence in law of X_∞ and $\bar{\Pi}_0(\tilde{H}_0^{-\infty})$ (Theorem 6.3a) implies $R_\infty(A) = R_{-\infty,0}(\bar{\Pi}_0^{-1}(A))$ and hence (6.17) implies

$$v_r = R_\infty(\{\mu:\mu(B(0,r))>0\}) = r^{d-\alpha/\beta}R_\infty(\{\mu:\mu(B(0,1))>0\}) = r^{d-\alpha/\beta}v_1,$$

where $c=v_1 \in [0,\infty]$. $c=0$ would imply $R_\infty=0$ and hence $X_\infty=0$ a.s. contradicting the persistence established in Theorem 6.1. The fact that $\alpha=2$ implies $c<\infty$ is proved in Corollary A.8.

(b) If $\phi \in bp\mathcal{B}(\mathbb{R}^d)$ and supp $\phi \subset B(0,r)$ then

$$\begin{aligned}-\log E(\exp\{-\langle X_\infty,\phi\rangle\}) &= \int(1-e^{-\langle\mu,\phi\rangle})R_\infty(d\mu)\\ &= \int \mathbf{1}(\mu(B(0,r))>0)(1-e^{-\langle\mu,\phi\rangle})R_\infty(d\mu)\\ &= v_r\int(1-e^{-\langle\mu,\phi\rangle})P_r^*(d\mu) \quad \text{(by (a))}.\end{aligned}$$

Finally the equivalence of the canonical measures in (a) and (6.12c) shows

$$\int\left(1 - e^{-\langle\mu,\phi\rangle}\right)R_\infty(d\mu) = \int\left(1 - e^{-\langle\mu,\phi\circ\Pi_0\rangle}\right)R_{-\infty,0}(d\mu)$$

$$= \bar{V}_0(\phi\circ\Pi_0). \tag{6.25}$$

Equating the right sides of the above expressions we obtain (6.24). ∎

Corollary 6.11. The localized clan measure given by (6.23) is $2/\beta$ self-similar, that is, $r^{-2/\beta}\langle X^{(r)},\phi(\cdot/r)\rangle$ is equal in law to $\langle X^{(1)},\phi\rangle$, if supp $\phi \subset B(0,1)$, and $X^{(r)}$ has law P_r^* given by (6.23).

Proof. This follows from (6.24), (6.22) and (6.19). ∎

Remarks.

(1) Consider the case $\alpha=2$, $\beta=1$, $d \geq 3$. For $t > 0$, $\varepsilon > 0$, let $X_t^\varepsilon(A) = \dfrac{X_t(\varepsilon A)}{\varepsilon^2}$, $A \subset B(0;r)$. It can be shown that conditioned on $X_t^\varepsilon(B(0;r)) \geq 0$, X_t^ε converges weakly as $\varepsilon \to 0$ to $\Pi^{(r)}$ (unpublished joint work with Iscoe.) The analogous result for X_∞ follows from Proposition 6.10(b). Note the contrast with

$$\overline{\lim_{\varepsilon \downarrow 0}} \frac{X_t(B(x,\varepsilon))}{\varepsilon^2 \log\log 1/\varepsilon} = c_1(2,d)$$

X_t-a.a. x Q_m-a.s. (Theorem 5.5(a)).

(2) R_∞ is an infinite measure on $M_p(\mathbb{R}^d)$ and in fact coincides with the infinite entrance law of Dynkin (1989) (also cf. El Karoui and Roelly-Coppoletta (1988)). The process ξ_t is the analogue of the "family dynamics" in Liemant, Matthes and Wakolbinger (1988). It describes the evolution of a clan which is a population of related individuals. The fact that R_∞ is an (infinite) invariant measure for ξ_t and the point process Ξ_t is invariant under the family dynamics is analogous to the fact that Lebesgue measure and the Poisson point process are invariant under the motion of Brownian particles. The analogous result for the case of discrete time is given in Liemant, Matthes and Wakolbinger (1988, Theorem 2.7.24).

(3) Actually the α/β-self-similarity of the Palm measure (Theorem 6.7) and the α/β-self-similarity of the canonical measure (Theorem 6.6(b)) are equivalent (cf. Zähle (1988a, 2.3)).

7. Weak Convergence of Branching Particle Systems

Our goal in this section is to extend the well-known construction of a superprocess as a weak limit of a system of branching Markov processes to the context of the historical process. We restrict ourselves to finite variance branching mechanisms $\Phi(x,\lambda) = -\sigma^2\lambda^2/2$. The spatial motions are given by $Y = (\Omega^2, \mathcal{F}^0, \mathcal{F}^0_{t+}, Y_t, P^x)$ which is assumed to be a Borel right process with càdlàg paths, conservative semigroup P_t and a Polish state space E with Borel σ-field $\mathcal{E}$. If Y is a Feller process on a separable locally compact state space the weak convergence result we will prove is well-known (see e.g. Ethier-Kurtz (1986, Sec. 9.4)). The historical process was constructed by a trivial (and continuous) projection of an ordinary superprocess but even in the context of super-Brownian motion this superprocess will not be a Feller process on a locally compact state space. Hence our main task is to extend the known weak convergence results to the more general spatial motions considered recently by Dynkin (1989a,b) and Fitzsimmons (1988). This extension, which will require some new methodology, is also of some intrinsic interest since much of our intuition about superprocesses comes from the approximating systems of branching particles.

Let U^α denote the α-resolvent of Y, B denote the class of bounded Borel-measurable, finely continuous functions on E, and, following Fitzsimmons (1988), define a weak generator of Y, $\underline{A}$, on $D(\underline{A}) = U^1(B)$ by $\underline{A}(U^1 g) = U^1 g - g$. The following elementary result is a routine exercise. $\xrightarrow{bp}$ denotes bounded pointwise convergence.

Lemma 7.1. If $f\in D(\underline{A})$ then $(P_t f-f)t^{-1} \xrightarrow{bp} \underline{A}f$ as $t\to 0+$ and

$$M_t^f = f(Y_t) - f(Y_0) - \int_0^t \underline{A}f(Y_s)ds$$

is an $(\mathcal{F}_{t+}^0)$-martingale w.r.t. any P^x.

We introduce some notation to describe the approximating systems of branching particles (see also Perkins (1988)). Fix an offspring distribution, ν, on $\mathbb{Z}^+$ with mean 1 and variance σ^2. To label the "branches" of the system of particles, introduce an index set $I = \bigcup_{n=0}^{\infty} \mathbb{N}^{n+1}$. If $(\alpha_0,\ldots,\alpha_n)\in I$, let $|\alpha| = n$, $\alpha|i = (\alpha_0,\ldots,\alpha_i)$ for $i\le n$ and write $\beta<\alpha$ if $\beta=\alpha|i$ for some $i\le n$. Let $I_n = \{\alpha\in I:|\alpha|<n\}$ $(n\in \mathbb{Z}_+)$ and $J_n = I_n-I_{n-1}$ $(n\in\mathbb{N})$. Let $E_\Delta = E\cup\{\Delta\}$ where Δ is added as an isolated point, extend $f:E\to\mathbb{R}$ to E_Δ by setting $f(\Delta) = 0$, and assume $Y_t\equiv \Delta$ under P^Δ. Let $\Omega_1 = E_\Delta^{\mathbb{N}}$, $\mathcal{F}_1 = \mathcal{B}(\Omega_1)$ and let $\{P_1^M:M\in\mathbb{N}\}$ be a sequence of probabilities on $(\Omega_1,\mathcal{F}_1)$ which will govern the initial configuration of particles. Let $\Omega=\Omega_1\times(D(E_\Delta)\times\mathbb{Z}_+)^I$, $\mathcal{F}$ denote the natural product σ-field and denote points in Ω by $\omega = ((x_i),(Y^\alpha,N^\alpha)_{\alpha\in I})$.

We will construct a probability P_M on $(\Omega,\mathcal{F})$ such that if $\omega = ((x_i),(Y^\alpha,N^\alpha))$, particles start from those $x_i\ne \Delta$ and on $[j/M,(j+1)/M)$ follow independent copies of Y, $\{Y_\cdot^\alpha:|\alpha|=j\}$. At $t = (j+1)/M$ each Y^α dies and branches, independently of each other, into N^α new particles, where N^α has law ν. The approximating measure-valued processes, $X^M(t,\omega)$, will assign mass

M^{-1} to the location of each particle alive at time t. The technical details follow.

Define an increasing sequence of σ-fields on Ω by

$$\mathcal{G}_n = \sigma(\{(x_i),(Y^\alpha,N^\alpha):\alpha\in I_n\}),\ n\in \mathbb{Z}_+.$$

By the Kolmogorov Extension Theorem for each $M\in \mathbb{N}$ there is a unique probability $P = P_M$ on $(\Omega,\mathcal{F})$ such that

$$P_M((x_i)\in A) = P_1^M(A)$$

$$(7.1)\quad P_M((Y^\alpha,N^\alpha)_{\alpha\in J_1}\in \prod_{\alpha\in J_1} A_\alpha\times B_\alpha|\mathcal{G}_0) = \prod_{\alpha\in J_1} P^{x_{\alpha_0}}(Y(.\wedge M^{-1})\in A_\alpha) \prod_{\alpha\in J_1} \nu(B_\alpha)$$

$$P_M((Y^\alpha,N^\alpha)_{\alpha\in J_{n+1}}\in \prod_{\alpha\in J_{n+1}} A_\alpha\times B_\alpha|\mathcal{G}_n)(\omega)$$

$$= \prod_{\alpha\in J_{n+1}} P^{x_{\alpha_0}}(Y(.\wedge(n+1)M^{-1})\in A_\alpha|Y(.\wedge nM^{-1})=Y^{\alpha|n-1}(\omega)) \prod_{\alpha\in J_{n+1}} \nu(B_\alpha).$$

Here $A_\alpha\in \mathcal{B}(D(E_\Delta))$ and $B_\alpha\subset \mathbb{Z}_+$. It follows that

$$(7.2)\quad P_M(Y^\alpha\in A|\mathcal{G}_0) = P^{x_{\alpha_0}}(Y(.\wedge(|\alpha|+1)M^{-1})\in A).$$

To prune this tree of branching Markov processes $\{Y^\alpha:\alpha\in I\}$ introduce death times

$$\tau^\alpha(\omega) = \begin{cases} 0 & \text{if } x_{\alpha_0}=\Delta \\ \min\{(i+1)/M:\ i\le|\alpha|,\ N^{\alpha|i} < \alpha_{i+1}\} & \text{if } x_{\alpha_0}\ne\Delta \text{ and this set is non-empty} \\ (|\alpha|+1)/M & \text{otherwise} \end{cases}$$

and define

$$X^\alpha(t,\omega) = \begin{cases} Y^\alpha(t,\omega) & \text{if } t < \tau^\alpha(\omega) \\ \Delta & \text{if } t \ge \tau^\alpha(\omega) \end{cases}$$

Write $\alpha \overset{M}{\sim} t$ (or $\alpha \sim t$ if there is no ambiguity) if

[illegible]

Define a random measure on $(E,\mathcal{E})$ by

$$X^M(t,\omega)(A) = M^{-1}\text{card}\{X_t^\alpha \in A:\ \alpha \sim t\}.$$

We will approximate the $(Y,-\sigma^2\lambda^2/2)$-historical process by the random measures on path space given by

$$H^M(t,\omega) = M^{-1}\sum_{\alpha\sim t} \delta_{(X_\cdot^\alpha)^t} \in M_F(D(E)).$$

Let $\mathcal{N}_M$ be the σ-field on Ω generated by the P_M-null sets and introduce the filtration

$$\mathcal{F}_t = \mathcal{F}_t^M = \sigma((x_i),Y^\alpha,N^\alpha:|\alpha|<i)\vee\Big(\bigcap_{u>t}\sigma(Y_s^\alpha:|\alpha|=i,s\le u)\Big)\vee\mathcal{N}_M$$

$$\text{if} \quad i/M\le t<(i+1)/M.$$

Then X^M and H^M are $(\mathcal{F}_t^M)$-adapted.

Turning to the initial measures, let us fix a non-zero initial m_0 in $M_F(E)$, Note that

$$(7.3) \qquad X^M(0) = \sum_{i=1}^\infty \delta_{x_i} 1(x_i\ne\Delta).$$

Let $m_0^M(\cdot) = P_M(X^M(0)(\cdot))$. To ensure convergence of $X_M(0)$ to m_0 we introduce two potential hypotheses on $\{P_1^M\}$:

$(H_{7.1})$ $P_1^M = \delta_{(x_i^M)_{i\in\mathbb{N}}}$ where $x_i^M \ne \Delta$ iff $i\le K_M$ and

$$M^{-1}\sum_{i=1}^{K_M} \delta_{x_i^M} \longrightarrow m_0 \quad \text{in } M_F(E) \quad \text{as } M\to\infty.$$

$(H_{7.2})$ Under P_1^M, $\{x_i:i\le K_M\}$ are i.i.d. random variables with law $m_0/m_0(E)$, $x_i=\Delta$ a.s. for $i>K_M$ and $\lim_{M\to\infty} K_M/M= m_0(E)$.

<u>Lemma 7.2.</u> Either $(H_{7.1})$ or $(H_{7.2})$ implies

$(H_{7.3})$ $m_0^M \longrightarrow m_0$ in $M_F(E)$, $X_0^M \xrightarrow{w} m_0$, and $\sup_M P_M(X^M(0)(E)^2) < \infty$.

<u>Proof.</u> Under $(H_{7.1})$ this is obvious (see (7.3)). Under $(H_{7.2})$

the weak convergence is an obvious consequence of the law of large numbers and the other conclusions are equally trivial. ∎

Lemma 7.3 (a) If $f \in p\mathcal{E}$, then $P_M(\langle X^M(t),f\rangle) = m_0^M P_t f$.

(b) If $f \in p\mathcal{B}(D(E))$, then $P_M(\langle H^M(t),f\rangle) = P^{m_0^M}(f(Y^t))$.

Proof. $$P_M(\langle X_t^M,f\rangle|\mathcal{G}_0) = M^{-1}\sum_{i=1}^{\infty}\sum_{\substack{\alpha\sim t\\ \alpha_0=i}} 1(x_i\neq\Delta)P_M(f(X_t^\alpha)|\mathcal{G}_0)$$

$$= M^{-1}\sum_{i=1}^{\infty}\sum_{\substack{\alpha\sim t\\ \alpha_0=i}} 1(x_i\neq\Delta)P_M(f(Y_t^\alpha)|\mathcal{G}_0)P_M(X_t^\alpha\neq\Delta|\mathcal{G}_0)$$

((7.1) implies Y^α and N^α are conditionally independent given $\mathcal{G}_0$)

$$= M^{-1}\sum_{i=1}^{\infty} P^{x_i}(f(Y_t))P_M\Big(\sum_{\substack{\alpha\sim t\\ \alpha_0=i}} 1(X_t^\alpha\neq\Delta)|\mathcal{G}_0\Big) \quad \text{(by (7.2))}$$

$$= \int P^x(f(Y_t))X_0^M(dx).$$

(a) follows upon taking expectations and (b) is similar. ∎

Notation. $T^M = \{i/M: i\in\mathbb{Z}_+\}$ and $\underline{t} = [tM]/M$ $(t\geq 0)$

If $Z: T^M\times\Omega \to \mathbb{R}$, let $\Delta Z(\underline{t}) = Z(\underline{t}+M^{-1}) - Z(\underline{t})$,

$$\langle Z\rangle(\underline{t}) = \sum_{0\leq\underline{s}<\underline{t}} P_M(\Delta Z(s)^2|\mathcal{F}_{\underline{s}}^M) + P_M(Z(0)^2),$$

$$\int_0^{\underline{t}} Z(\underline{s})d\underline{s} = \sum_{0\leq\underline{s}<\underline{t}} Z(\underline{s})M^{-1}.$$

Lemma 7.4. Assume $(H_{7.3})$. If $f\in D(\underline{A})$, then

$$\langle X_t^M,f\rangle = \langle X_0^M,f\rangle + \int_0^t X_s^M(\underline{A}f)ds + Z_t^M(f) + E_t^M(f), \tag{7.4}$$

where

(a) $\{\langle X_\cdot^M,f\rangle: M\in\mathbb{N}\}$ is tight in $D(\mathbb{R})$ and all weak limit points are continuous.

(b) $Z_t^M(f) = Z_{\underline{t}}^M(f)$ is a discrete-time $\{\mathcal{F}_{\underline{t}}^M : \underline{t} \in T^M\}$-martingale under P_M such that

(i) $\{Z_{\cdot}^M(t) : M \in \mathbb{N}\}$ is tight in $D(\mathbb{R})$ and all limit points are continuous,

(ii) $\langle Z^M(f)\rangle(\underline{t}) = \sigma^2 \int_0^{\underline{t}} \langle X_{\underline{s}}^M, f^2\rangle d\underline{s}$

(iii) $\{\sup_{\underline{t} \le T} Z_{\underline{t}}^M(f)^2 : M \in \mathbb{N}\}$ is uniformly integrable for any $T > 0$

(P_M is the underlying probability for Z^M).

(c) $\lim_{M\to\infty} P_M(\sup_{t \le T} E_t^M(f)^2) = 0$ for any $T > 0$.

<u>Proof.</u> Fix M and write P for P_M and Δt for M^{-1}. Let $f \in D(\underline{A})$. If $\alpha \sim \underline{r}$ and $s \in (\underline{r}, \underline{r}+\Delta t]$, let

$$M_{\underline{r},s}^{\alpha} = \begin{cases} f(Y_s^\alpha) - f(Y_{\underline{r}}^\alpha) - \int_{\underline{r}}^{s} \underline{A} f(Y_v^\alpha)\,dv & \text{if } X_{\underline{r}}^\alpha \ne \Delta \\ 0 & \text{if } X_{\underline{r}}^\alpha = \Delta \end{cases}.$$

Define

$$Z_{\underline{s}}^M(f) = Z_{\underline{s}}(f) = M^{-1} \sum_{\underline{r} < \underline{s}} \sum_{\alpha \sim \underline{r}} f(X_{\underline{r}}^\alpha)(N^\alpha - 1)$$

$$N_{\underline{s}}(f) = M^{-1} \sum_{\underline{r} < \underline{s}} \sum_{\alpha \sim \underline{r}} \int_{\underline{r}}^{\underline{r}+\Delta t} \underline{A} f(X_r^\alpha)\,dr\,(N^\alpha - 1)$$

$$M_s(f) = M^{-1} \sum_{\underline{r} < \underline{s}} \sum_{\alpha \sim \underline{r}} M_{\underline{r},\underline{r}+\Delta t}^\alpha N^\alpha + M^{-1} \sum_{\alpha \sim \underline{s}} M_{\underline{s},s}^\alpha ,$$

and extend $Z_{\underline{s}}(f)$ and $N_{\underline{s}}(f)$ to $[0,\infty)$ as right continuous step functions. Then

$$\langle X_{\underline{s}+\Delta t}^M, f\rangle - \langle X_{\underline{s}}^M, f\rangle = M^{-1} \sum_{\alpha \sim \underline{s}} (f(Y_{\underline{s}+\Delta t}^\alpha) N^\alpha - f(Y_{\underline{s}}^\alpha)) 1(X_{\underline{s}}^\alpha \ne \Delta)$$

$$= M^{-1} \sum_{\alpha \sim \underline{s}} (M_{\underline{s},\underline{s}+\Delta t}^\alpha + \int_{\underline{s}}^{\underline{s}+\Delta t} \underline{A} f(Y_r^\alpha)\,dr) N^\alpha 1(X_{\underline{s}}^\alpha \ne \Delta)$$

$$+ \quad M^{-1} \sum_{\alpha\sim\underline{s}} f(X^{\alpha}_{\underline{s}})(N^{\alpha}-1)$$

$$(7.5) \qquad = \Delta M(f)(\underline{s}) + \Delta N(f)(\underline{s}) + M^{-1} \sum_{\alpha\sim\underline{s}} \int_{\underline{s}}^{\underline{s}+\Delta t} \underline{A} f(X^{\alpha}_{r})dr + \Delta Z(f)(\underline{s}).$$

Also

$$(7.6) \quad \langle X^M_s, f\rangle - \langle X^M_{\underline{s}}, f\rangle = M^{-1} \sum_{\alpha\sim\underline{s}} M^{\alpha}_{\underline{s},s} + M^{-1} \sum_{\alpha\sim\underline{s}} \int_{\underline{s}}^{s} \underline{A} f(X^{\alpha}_{r})dr.$$

(7.5) and (7.6) together imply (7.4) where $E^M_t(f) = M_t(f) + N_t(f)$. Lemma 7.1 and (7.1) imply that if $\alpha\sim\underline{r}$, then $\{(M^{\alpha}_{\underline{r},s}, \mathcal{F}_s): s\in[\underline{r},\underline{r}+\Delta t]\}$ is a martingale. (7.1) also shows

(7.7) $\{M^{\alpha}_{\underline{r},\cdot}, N^{\alpha}: \alpha\sim\underline{r}\}$ are conditionally independent given $\mathcal{F}_{\underline{r}}$.

Therefore $M_s(f)$ is an $(\mathcal{F}_s)$-martingale. If

$$H(\delta,y) = P^y\left(\left(f(Y_\delta)-f(y) - \int_0^\delta \underline{A} f(Y_r)dr\right)^2\right),$$

then

$$(7.8) \quad H(\delta,y) \le 2P^y((f(Y_\delta)-f(y))^2) + 2\delta^2\|\underline{A}f\|^2 \xrightarrow{bp} 0 \quad \text{as} \quad \delta \to 0$$

by the fine continuity of f. If $T\in\mathbb{N}$, then

$$P_M(\langle M(f)\rangle_T) = (\sigma^2+1)M^{-2}P_M\left(\sum_{\underline{r}<T} \sum_{\alpha\sim\underline{r}} P_M((M^{\alpha}_{\underline{r},\underline{r}+\Delta t})^2|\mathcal{F}_{\underline{r}})1(X^{\alpha}_{\underline{r}}\ne\Delta)\right)$$

(by (7.7)

$$= (\sigma^2+1)P_M\left(\sum_{\underline{r}<T} \sum_{\alpha\sim\underline{r}} H(\Delta t, X^{\alpha}_{\underline{r}})M^{-2}\right)$$

$$= (\sigma^2+1) \int_0^T m^M_0 P_{\underline{r}}(H(\Delta t,\cdot))d\underline{r} \qquad \text{(Lemma 7.3)}$$

Doob's maximal inequality and (7.8) imply

$$(7.9) \quad \lim_{M\to\infty} P_M\left(\sup_{s\le T} M_s(f)^2\right) = 0 \quad \text{for all } T\in\mathbb{N}.$$

(7.7) and $P_M(N^\alpha) = 1$ show that $\{(N_{\underline{s}}(f),\mathcal{F}_{\underline{s}}):\underline{s}\in T_M\}$ is a martingale. An argument similar to (but simpler than) the above shows

(7.10) $$\lim_{M\to\infty} P_M(\sup_{s\le T} N_s(f)^2) = 0 \quad \text{for all } T\in\mathbb{N}.$$

(7.9) and (7.10) imply (c).

The proof of (b) is identical to that in the well-known Feller setting (see Ethier-Kurtz (1986, Sec. 9.4)) - our deterministic branching leads only to minor changes. (a) is an easy consequence of (7.4), (b) and (c). ∎

As the above results suggest, to prove X_M converges weakly to the $(Y,-\sigma^2\lambda^2/2)$-superprocess we will use the following martingale characterization of the superprocess, which is a special case of Fitzsimmons (1988, Theorem 4.1), (1989). Recall from Theorem 2.1.3 that $X = (\Omega^0,\mathcal{G}^0,\mathcal{G}^0_{t+},\theta_t,X_t,Q_m)$ denotes the (Y,Φ)-superprocess on the space of càdlàg $M_F(E)$-valued paths.

<u>Theorem</u> <u>7.5.</u> Assume $\Phi(x,\lambda) = -\sigma^2\lambda^2/2$ and $m\in M_F(E)$. Then Q_m is the unique probability on $(\Omega^0,\mathcal{G}^0)$ such that $Q_m(X_0=m) = 1$ and for any $f\in \mathcal{D}(\underline{A})$

$$Z_t(f) = X_t(f) - m(f) - \int_0^t X_s(\underline{A}f)ds$$

is a continuous $\mathcal{G}^0_{t+}$-martingale such that $\langle Z(f)\rangle_t = \int_0^t \sigma^2 X_s(f^2)ds$. Moreover Q_m is supported by the continuous $M_F(E)$-valued paths.

If Y is a Feller process on a separable locally compact space the (well-known) weak convergence of $\{X^M\}$ to Q_{m_0} is an easy consequence of Lemma 7.4 and Theorem 7.5. There are two problems to face when extending this result to the current setting. The first is a compactness problem due to the weaker hypothesis on E (which is essential to accomodate the historical process). The second problem arises because the "test functions" in $D(\underline{A})$ need not be continuous in general and hence will not mix well with the weak topology on $M_F(E)$. One solution for this latter problem would be to work with the Ray topology but we shall see that this will not be necessary if Y is a Hunt process. We first consider the compactness problem. This is essentially solved in Proposition 7.7 by using a type of "Reflection Principle" on the approximating historical processes.

<u>Notation.</u> If $w \in D(E)$ define $w^{t-} \in D(E)$ by

$$w^{t-}(u) = \begin{cases} w(u) & \text{if } u < t, \\ w(t-) & \text{if } u \geq t \end{cases}$$

where $w(0-) = w(0)$. If $A \subset D(E)$ let

$$A_\infty = \{w^t, w^{t-} : w \in A, t \in [0,\infty]\} \quad (w^\infty = w^{\infty-} = w).$$

An easy calculation shows

<u>Lemma 7.6.</u> Assume $K \subset D(E)$ is compact. Then K_∞ is also compact in $D(E)$, and

$$V(K,N) = \{w(t), w(t-) : t \leq N, w \in K\}$$

is compact in E for all $N \geq 0$.

<u>Proposition 7.7.</u> For any $\varepsilon > 0$ there is a $\delta > 0$ such that if $K \subset D(E)$ is compact and

$$(7.11) \quad \sup_M P^{m_0^M}(Y \in K^c) \leq \delta \text{ and } \sup_M m_0^M(E) \leq \varepsilon^{-1},$$

then

$$\text{(a)} \quad \sup_M P_M(H_t^M(K_\infty^c) > \varepsilon \text{ for some } t \geq 0) \leq \varepsilon,$$

and

$$\text{(b)} \quad \sup_M P_M(X_t^M(V(K,\delta^{-1})^c) > \varepsilon \text{ for some } t \geq 0) \leq \varepsilon.$$

<u>Proof.</u> Let $\varepsilon > 0$ and $\delta > 0$. Assume (7.11). By the Section

Theorem there is an $(\mathcal{F}_t^M)$-stopping time $S = S_M$ such that

(7.12) $[S] \subset \{(t,\omega): H_t^M(K_\infty^c) \geq \varepsilon)$,

$$P_M(S<\infty) > P(H_t^M(K_\infty^c) > \varepsilon \text{ for some } t\geq 0) - \varepsilon/2.$$

Let $\{Z^M(t): t\in T^M\}$ denote the Markov chain with state space T^M and whose transition function is that of $X_t^M(E)$. Hence $Z^M(t)$ is a rescaled Galton-Watson process which by Feller's Theorem (see Ethier-Kurtz (1986, Ch. 9, Thm. 1.3)) converges weakly to the unique solution of $Z_t = z_0 + \int_0^t \sigma(Z_s)^{1/2} dB_s$ (B is a Brownian motion). We abuse the notation and use P^x to denote probabilities for both Z^M and Z starting at x (not to mention Y!). If

$$p(M,\varepsilon',u) = P^{(1+[\varepsilon' M])/M}(\inf_{t\leq u} Z_t^M > \varepsilon'/2),$$

then

$$(7.13)\quad \lim_{M\to\infty} p(M,\varepsilon',u) = p(\varepsilon',u) \equiv P^{\varepsilon'}(\inf_{t\leq u} Z_t > \varepsilon'/2) > 0$$

for all $\varepsilon',u > 0$.

Let $K^t = \{w^t: w\in K\}$, and on $\{S<\infty\}$ define $\tilde{S} = [MS]/M$ and

$$\hat{Z}_t = \begin{cases} H^M_{\tilde{S}+t}(w: w^S \notin K^S\}) & \text{if } t \in T^M - \{0\} \\ H_S^M(\{w: w^S \notin K^S\}) & \text{if } t=0 \end{cases}$$

The strong Markov property for H^M and the independence of the branching variables $\{N^\alpha\}$ and the spatial motions $\{Y^\alpha\}$ shows that on $\{S<\infty\}$

$$(7.14)\quad P_M(\hat{Z}\in A|\mathcal{F}_S)(\omega) = P^{H_S^M((K^S)^c)(\omega)}(Z^M\in A)\ P_M\text{-a.s. for } A \subset (T^M)^{T^M}.$$

Note that on $\{S<u\}$, $w^S \notin K^S$ implies $w \notin K^u$ and so on $\{S<u\}$ $(u\in T^M)$

$$H_u^M((K^u)^c) \geq H_u^M(\{w: w^S \notin K^S\}) = \hat{Z}_{u-\tilde{S}}$$

and therefore by (7.14)

$$P_M(S<u,\ H_u^M((K^u)^c)\le \varepsilon/2) \le P_M(1(S<u)P_M(\hat{Z}_{u-\tilde{S}} \le \varepsilon/2|\mathcal{F}_S))$$

$$\le P_M(1(S<u)P^{H_S^M((K^S)^c)}(\inf_{t\le u} Z_t^M \le \varepsilon/2))$$

$$\le P_M(S<u)(1-p(M,\varepsilon,u))$$

because $H_S^M((K^S)^c) \ge H_S^M(K_\infty^c) \ge \varepsilon$ on $\{S<\infty\}$. This shows that for $u\in T^M$,

$$p(M,\varepsilon,u)P_M(S<u) \le P_M(S<u,H_u^M((K^u)^c) > \varepsilon/2)$$

$$\le 2\varepsilon^{-1}P_M(H_u^M((K^u)^c))$$

$$= 2\varepsilon^{-1}P_M^{m_0^M}(Y^u\in (K^u)^c) \quad \text{(Lemma 7.3(b))}$$

$$\le 2\varepsilon^{-1}P^{m_0^M}(Y\notin K) \le 2\varepsilon^{-1}\delta.$$

(7.12) and the above give

$$P_M(H_t^M(K_\infty^c) > \varepsilon \text{ for some } t<\infty) \le P_M(S<u) + P_M(u\le S<\infty) + \varepsilon/2$$

(7.15) $$\le 2(p(M,\varepsilon,u)\varepsilon)^{-1}\delta + P_M(X_u^M(E)>0) + \varepsilon/2$$

$$= 2(p(M,\varepsilon,u)\varepsilon)^{-1}\delta + P_M(q_M(X_0^M(E),u)) + \varepsilon/2,$$

where

$$q_M(x,u) = P^{[Mx]/M}(Z^M([uM]/M) > 0).$$

The second part of (7.11) implies that $P_M(X_0^M(E)>N) \le (N\varepsilon)^{-1}$ and so (7.15) implies that for $u\in T^M$, $N>0$

(7.16) $$P_M(H_t^M(K_\infty^c)>\varepsilon \text{ for some } t\ge 0)$$
$$\le 2(p(M,\varepsilon,u)\varepsilon)^{-1}\delta + q_M(N,u) + (\varepsilon N)^{-1} + \varepsilon/2.$$

Feller's Theorem implies

$$\lim_{M\to\infty} q_M(N,u) = P^N(Z_u>0) = 1-e^{-2N/u\sigma^2} \quad \text{(Knight (1981, p. 100))}$$

$$\longrightarrow 0 \text{ as } u\to\infty$$

and therefore

(7.17) $\lim_{u\to\infty} \sup_{M\in\mathbb{N}} q^M(N,u) = 0$ for any $N>0$.

It also follows from (7.13) that

(7.18) $\inf_{M\in\mathbb{N}} p(M,\varepsilon,u) > 0$ for any $u>0$.

By (7.17) and (7.18) we may first choose N and then $u\in\mathbb{N}$ sufficiently large and then $\delta=\delta(\varepsilon)$ sufficiently small so that the right-hand side of (7.16) is less than ε for all $M\in\mathbb{N}$. This completes the proof of (a).

For (b) note that $V(K,u) = V(K_\infty,u)$ and so if $t\le u$ then

$$X_t^M(V(K,u)^c) = H_t^M(\{w:w_t\notin V(K_\infty,u)\}) \le H_t^M(K_\infty^c).$$

Therefore

$$P_M(X_t^M(V(K,u)^c) > \varepsilon \text{ for some } t\ge 0)$$

$$\le P_M(H_t^M(K_\infty^c)>\varepsilon \text{ for some } t\in[0,u]) + P_M(X_u^M(E)>0)$$

(7.19) $$\le \varepsilon + q_M(N,u) + (\varepsilon N)^{-1},$$

as in the proof of (a). As before we may first choose N and then u sufficiently large so that the right-hand side of (7.19) is less than 2ε for all $M\in\mathbb{N}$. By making δ smaller, if necessary, we may assume $\delta^{-1} \ge u$ ($u=u(\varepsilon)$ chosen above) and (b) follows. ■

<u>Corollary 7.8</u> Assume $\{P^{m_0^M}(Y\in\cdot): M\in\mathbb{N}\}$ is relatively compact in $M_F(D(E))$. For any $\varepsilon>0$ there are compact sets K_0, K_1 in $D(E)$ and E, respectively, such that

$$\sup_{M\in\mathbb{N}} P_M(H_t^M(K_0^c) > \varepsilon \text{ for some } t \ge 0) \le \varepsilon,$$

and

$$\sup_{M\in\mathbb{N}} P_M(X_t^M(K_1^c) > \varepsilon \text{ for some } t\ge 0) \le \varepsilon.$$

<u>Proof.</u> This is immediate from Proposition 7.7 and Lemma 7.6. ■

We now turn to the second obstacle, namely, the potential discontinuities of functions in $D(\underline{A})$. Our goal is to construct "large" compact sets on which functions in $D(\underline{A})$ are continuous.

To do this we use the following version of Lusin's Theorem.

Proposition 7.9. Let X be compact metric, Y separable metric and $f:X \to Y$ be Borel measurable. If $\nu \in M_F(X)$, then for any $\delta > 0$ there is a compact set K in X such that $\nu(K^c) < \delta$ and $f|_K$ is continuous.

Proof. The well-known proof for $Y=\mathbb{R}$ (e.g. Halmos (1950, p.243)) extends without difficulty. ∎

The problem is that our compact sets must be simultaneously large with respect to the infinite collection of random measures $\{X_t^M : M \in \mathbb{N}, t \geq 0\}$. If we do not assume $(H_{7.2})$ this problem poses unsurmountable difficulties (see Theorem 7.16(b)), unless of course we impose continuity conditions on $U^\alpha f$ (see Theorem 7.16(a)). If $f \in \mathcal{D}(\underline{A})$ and Y is a Hunt process then the facts that $f \circ Y$ is càdlàg and $f(Y(t-)) = f \circ Y(t-)$ will allow us to use Lusin's Theorem to find a compact set in $D(E)$ which is large for $P^m(Y \in .)$ and on which $y \to f(y(t))$ and $y \to f(y(t-))$ are continuous uniformly in t (Lemma 7.10). Proposition 7.7 will then show that K is "large" with respect to the infinite collection of random measures mentioned above (Lemma 7.11). $(H_{7.2})$ will be needed to verify hypothesis (7.11) of Proposition 7.7.

We have sketched our line of reasoning because the statements and proofs of Lemmas 7.10 and 7.11, and Theorem 7.12 use nonstandard analysis. The main weak convergence results (Theorems 7.13 and 7.14) are then stated in standard terms, and are both elementary consequences of Theorem 7.12. Self-contained introductions to nonstandard analysis may be found in Loeb (1979) or Cutland (1983).

We work in an ω_1-saturated enlargement of a superstructure containing $\mathbb{R}$ and E. st_S denotes the standard part map on the nearstandard points $ns(*S)$ of a topological Hausdorff space S. If $S \subset \mathbb{R}^n$ we write $°x$ for $st_S(x)$. If ν is an internal measure on an internal algebra $\mathcal{A}$, $L(\nu)$ denotes the associated Loeb measure on $\sigma(\mathcal{A})$ and its completion $L(\mathcal{A})$.

<u>Lemma</u> <u>7.10</u> Assume Y is a Hunt process and $f = U^\alpha g$ for $g \in b\mathcal{E}$ and $\alpha > 0$. For any $\delta > 0$ and $m \in M_F(E)$ there is a compact $K \subset D(E)$ such that $P^m(Y \in K^c) < \delta$ and

(7.20) $°*f(y_t) = f(st_E(y_t))$ and $°*f(y_{t-}) = f(st_E(y_{t-}))$

for all $t \in ns(*[0,\infty))$ and $y \in *K$.

<u>Proof.</u> Let f be as above and

$$A = \{y \in D(E):\ f\circ y \in D(\mathbb{R}),\ f(y_{t-}) = f\circ y(t-) \text{ for all } t > 0\}.$$

It is easy to see that A^c is an analytic subset of $D(\mathbb{R})$. $f\circ Y \in D(\mathbb{R})$, P^x-a.s. for any x by Blumenthal and Getoor (1968, Ch. II, Thm. 2.1.2). Note that

$$e^{-\alpha t}U^\alpha g(Y_t) = E^x\left(\int_0^\infty e^{-\alpha u} g(Y_u)\,du \,|\, \mathcal{F}^0_{t+}\right) - \int_0^t e^{-\alpha u} g(Y_u)\,du.$$

By quasi-left-continuity the martingale part cannot jump at predictable times (Sharpe (1988, 47.6)) and, as the same is true of Y, $f(Y_{S-}) = f\circ Y(S-) = f(Y_S)$ for any predictable time S. The predictable section theorem implies $f(Y_{t-}) = f\circ Y(t-)$ for all $t \ge 0$, P^x-a.s. for any $x \in E$. We have shown $P^x(Y \in A) = 1$ for all x. If $\delta > 0$ the regularity of P^m implies there is a compact $K_0 \subset A$ such that $P^m(Y \in K_0^c) < \delta$.

Define $\tilde{f}: K_0 \to D(E\times\mathbb{R})$ by $\tilde{f}(y)(t) = (y(t), f\circ y(t))$. By Lusin's Theorem (Proposition 7.9) there is a compact set $K \subset K_0$ such that $P^m(Y \in K^c) < \delta$ and $\tilde{f}|_K$ is continuous. If $y \in {}^*K$ and

$y_0 = st_{D(E)}(y)$, the continuity of $\tilde{f}|_K$ implies $(y, {}^*f\circ y) \in ns({}^*D(E\times\mathbb{R}))$ and $st_{D(E\times\mathbb{R})}(y, {}^*f\circ y)(t) = (y_0(t), f\circ y_0(t))$ for $t \in [0,\infty)$. The characterization of $st_{D(E\times\mathbb{R})}$ in Hoover-Perkins (1983, Thm.2.6) shows that for any $t \in [0,\infty)$ there is a $t' \approx t$ $(0'=0)$ such that

$$(7.21) \quad st_{E\times\mathbb{R}}(y(s), {}^*f(y(s))) = \begin{cases} (y_0(t-), f\circ y_0(t-)) & \text{if } {}^\circ s=t,\ s<t' \\ (y_0(t), f\circ y_0(t)) & \text{if } {}^\circ s=t,\ s\geq t' \end{cases}$$

$$= \begin{cases} (y_0(t-), f(y_0(t-))) & \text{if } {}^\circ s=t,\ s<t' \\ (y_0(t), f(y_0(t))) & \text{if } {}^\circ s=t,\ s\geq t' \end{cases}$$

since $y_0 \in K \subset A$. Using the equality of the first components of (7.21) we have

$$(7.22) \quad f(st_E(y(s))) = \begin{cases} f(y_0(t-)) & \text{if } {}^\circ s=t,\ s<t' \\ f(y_0(t)) & \text{if } {}^\circ s=t,\ s\geq t' \end{cases}.$$

Compare this with the second components in (7.21) to see that $f(st_E(y(s))) = {}^\circ{}^*f(y(s))$ for all ${}^\circ s<\infty$. A similar argument in which s is replaced by $s-$ in the left side of (7.21) and (7.22) and $s<t'$ is replaced by $s\leq t'$ on the right side gives the result for left-hand limits. ∎

<u>Remark.</u> In standard terms, (7.20) states that the mapping $y \longrightarrow f(y(t))$ and $y \longrightarrow f(y(t-))$ are continuous on K uniformly for t in bounded subsets of $[0,\infty)$.

<u>Lemma 7.11.</u> Assume Y is a Hunt process and for some $m \in M_F(E)$

$$(7.23) \quad \sup_M P^{m_0^M}(Y\in A) \leq P^m(Y\in A) \quad \text{for all } A \in \mathcal{B}(D(E)).$$

If $f = U^\alpha g$ for some $g \in b\mathcal{E}$ and $\alpha>0$, then for any $M \in {}^*\mathbb{N}$,

$x \in ns({}^*E)$ and ${}^\circ{}^*f(x) = f(st_E(x))$

for $L(X_t^M)$-a.a. x and all $t \in ns({}^*[0,\infty))$ $L({}^*P_M)$-a.s.

<u>Proof.</u> Let ε be sufficiently small so that $\sup_M m_0^M(E) \leq \varepsilon^{-1}$. Choose $\delta = \delta(\varepsilon)$ as in Proposition 7.7 and for this δ and the

dominating m in (7.23), let $K \subset D(E)$ be the compact set obtained in Lemma 7.10. It is easy to see that (7.20) holds with K_∞ in place of K. Therefore if $M \in {}^*\mathbb{N}$ and $t \in ns({}^*[0,\infty))$,

$$L(X_t^M)(\{x \in {}^*E: {}^{\circ *}f(x) \neq f(st_E(x)) \text{ or } x \notin ns({}^*E)\})$$
$$= L(H_t^M)(\{y \in {}^*D(E): {}^{\circ *}f(y(t)) \neq f(st_E(y(t))) \text{ or } y(t) \notin ns({}^*E)\})$$
$$\leq L(H_t^M)({}^*K_\infty^c) = {}^\circ H_t^M({}^*K_\infty^c).$$

The choice of K and (7.23) shows that (7.11) holds and hence Proposition 7.7 and the above imply

$$\sup_{M \in {}^*\mathbb{N}} L({}^*P_M)(L(X_t^M)(\{x: {}^{\circ *}f(x) \neq f(st_E(x)) \text{ or } x \notin ns({}^*E)\}) > \varepsilon \text{ for some } t \geq 0)$$
$$\leq {}^\circ \sup_{M \in {}^*\mathbb{N}} {}^*P_M(H_t^M({}^*K_\infty^c) > \varepsilon \text{ for some } t \geq 0) \leq \varepsilon.$$

Let $\varepsilon \downarrow 0$ to complete the proof. ∎

<u>Notation.</u> If $C \subset b\mathcal{E}$, $\bar{C}^{bp}$ denotes the closure of C under bounded pointwise convergence.

<u>Theorem 7.12.</u> Assume $\{P^{m_0^M}(Y \in .): M \in \mathbb{N}\}$ is relatively compact in $M_F(D(E))$, $(H_{7.3})$ holds, and there is a $C \subset B$ such that $\bar{C}^{bp} = b\mathcal{E}$,

(7.24) $\langle X_0^M, U^1 f\rangle \xrightarrow{w} \langle m_0, U^1 f\rangle$ as $M \to \infty$, for all $f \in C$,

and for all $M \in {}^*\mathbb{N}-\mathbb{N}$, $f \in C$ and $\alpha > 0$,

(7.25) ${}^{\circ *}f(x) = f(st_E(x))$ and ${}^{\circ *}U^\alpha f(x) = U^\alpha f(st_E(x))$

for $L(X_t^M)$-a.a. x and all $t \in ns({}^*[0,\infty))$, $L({}^*P_M)$-a.s.

Then $\{X^M: M \in \mathbb{N}\}$ converges weakly in $D(M_F(E))$ to Q_{m_0}, the law of the $(Y, -\sigma^2\lambda^2/2)$-superprocess, starting at m_0. (Q_{m_0} is supported by $C(M_F(E))$). Moreover for any $f \in C$ and $\alpha > 0$,

(7.26) $\langle X^M, U^\alpha f\rangle \xrightarrow{w} Q_{m_0}(\langle X, U^\alpha f\rangle \in .)$ in $D(\mathbb{R})$

(where the limit is supported by $C(\mathbb{R})$).

<u>Proof.</u> Fix $M \in {}^*\mathbb{N}-\mathbb{N}$ and write P for $L({}^*P_M)$. Proposition 7.7 shows that $L(X_t^M)(ns({}^*E)^c) = 0$ for all $t \in {}^*[0,\infty)$ P-a.s.. Since

${}^\circ X_t^M({}^*E) < \infty$ for all ${}^\circ t < \infty$ P-a.s. (set f=1 in Lemma 7.4(a)) we may conclude that $X_t^M \in ns({}^*M_F(E))$ for all ${}^\circ t<\infty$ (see Anderson-Rashid (1978)), for ω outside a P-null set, N. Let $x_t = st_{M_F(E)}(X_t^M) = L(X_t^M)(st_E^{-1}(\cdot))$ for $t\in ns({}^*[0,\infty))$ and $\omega\in N^c$ and let $x_t \equiv 0$ on N. Let $\alpha > 0$ and

$$\mathcal{D} = \{f\in b\mathcal{E}: \langle x_t, U^\alpha f\rangle = \langle x_s, U^\alpha f\rangle \text{ for all } s\approx t \text{ in } ns({}^*[0,\infty)) \text{ P-a.s.}\}.$$

If $f\in \mathcal{C}$, fix $\omega\notin N$ and outside a P-null set so that (7.25) holds and $\langle X^M_\cdot, U^\alpha f\rangle$ is S-continuous (recall $U^\alpha(B) = U^1(B)$ and use Lemma 7.4(a)). If $s\approx t\in ns({}^*[0,\infty))$, then

$$\begin{aligned}\langle x_t, U^\alpha f\rangle = \langle L(X_t^M), U^\alpha f\circ st_E\rangle &= {}^\circ\langle X_t^M, {}^*U^\alpha f\rangle \quad \text{(by (7.25))}\\ &= {}^\circ\langle X_s^M, {}^*U^\alpha f\rangle \quad \text{(by S-continuity)}\\ &= \langle x_s, U^\alpha f\rangle.\end{aligned}$$

Therefore $\mathcal{C} \subset \mathcal{D}$ and since $\mathcal{D} = \bar{\mathcal{D}}^{bp}$, it follows that $\mathcal{D} = b\mathcal{E}$. If $\varepsilon>0$ we may choose a compact set $K_1 \subset E$ as in Corollary 7.8. Therefore

$$\begin{aligned}(7.27)\qquad & P(x_t(K_1^c)>\varepsilon \text{ for some } t\ge 0)\\ &= P(L(X_t^M)(st^{-1}(K_1^c))>\varepsilon \text{ for some } t\ge 0)\\ &\le {}^\circ{}^*P_M(X_t^M({}^*K_1^c)>\varepsilon \text{ for some } t\ge 0) \le \varepsilon.\end{aligned}$$

Let C' be a countable dense set in $C(K_1,\mathbb{R})$, extend each $f\in C'$ to a bounded continuous function on E (by Tietze), and denote this class of extensions by $C'' \subset C_b(E)$. Since $\alpha U^\alpha f \xrightarrow{bp} f$ as $\alpha\to\infty$ ($\alpha\in \mathbb{N}$, say) for $f\in C''$, the fact that $\langle x_s, U^\alpha f\rangle = \langle x_t, U^\alpha f\rangle$ for all $s\approx t$ in $ns({}^*[0,\infty))$, $\alpha\in\mathbb{N}$ and $f\in C''$ P-a.s. implies $x_s|_{K_1} = x_t|_{K_1}$ for all such s,t a.s. (7.27) therefore shows that $x_s=x_t$ for all $s\approx t$ in $ns({}^*[0,\infty))$ P-a.s. Therefore $X^M: ns({}^*[0,\infty)) \longrightarrow ns({}^*M_F(E))$ is S-continuous a.s. and we may define a continuous $M_F(E)$-valued process by $X(t) = st(X^M)(t)$ where st is the standard part map on ${}^*D(M_F(E))$ (define $X_t\equiv 0$ on the exceptional set where X^M is not

S-continuous).

Lemma 7.4 shows that for $f\in C$

(7.28) $\langle X^M_t, U^1f\rangle = \langle X^M_0, U^1f\rangle + \int_0^t \langle X^M_s, U^1f-f\rangle ds + Z^M_t(U^1f) + E^M_t(U^1f)$.

Let $\{\mathcal{F}_t : t\geq 0\}$ be the standard part of the internal filtration $\{{}^*\mathcal{F}^M_{\underline{t}} : \underline{t}\in T^M\}$ (see Hoover-Perkins (1983, Section 3)) and let $Z_t(U^1f) = st_{D(E)}(Z^M(U^1f))(t)$. Lemma 7.4(b) and Hoover-Perkins (1983, Theorems 5.2,6.7,8.5) imply $Z_t(U^1f)$ is a continuous $\mathcal{F}_t$-martingale such that

$$\langle Z(U^1f)\rangle_{{}^\circ\underline{t}} = {}^\circ\!\int_0^{\underline{t}} \sigma^2 \langle X^M_{\underline{s}}, (U^1f)^2\rangle d\underline{s} \quad \text{for } {}^\circ\underline{t} < \infty, \text{ a.s.}$$

Here Lemma 7.4(b)(iii) allows us to conclude $Z^M_t(U^1f)^2$ is S-integrable and hence apply Theorem 8.5(b) of Hoover-Perkins (1983). (7.25) and Anderson's construction of Lebesgue measure imply

(7.29) $\langle Z(U^1f)\rangle_t = \sigma^2 \int_0^t \langle X_s, (U^1f)^2\rangle ds.$

Take standard parts in (7.28). Use (7.24) to handle the initial condition, (7.25) to handle terms involving X^M and Lemma 7.4(c) to see that $E^M_\cdot(U^1f) \approx 0$ a.s. We conclude that

$$\langle X_t, U^1f\rangle = \langle m_0, U^1f\rangle + \int_0^t \langle X_s, U^1f-f\rangle ds + Z_t(U^1f)$$

for all $t\geq 0$, a.s.

(7.30)

$Z_t(U^1f)$ is a continuous $(\mathcal{F}_t)$-martingale

such that (7.29) holds.

Take bounded pointwise limits to see that (7.30) holds for all $f\in b\mathcal{E}$. By (7.30) $Z_t(U^1f)$ is also an $\mathcal{F}^X_{t+}$-martingale ($\mathcal{F}^X_t = \sigma(X_s, s\leq t)$). Theorem 7.5 implies that Q_{m_0} is the law of X. The nonstandard characterization of weak convergence

(Anderson-Rashid (1978)) shows that $X^M \xrightarrow{w} Q_{m_0}$. Finally (7.26) is a consequence of $st_{D(\mathbb{R})}(\langle X^M_{\cdot}, U^\alpha f\rangle)(t) = \langle X_t, U^\alpha f\rangle$, which is immediate from $X = st_{D(M_F(E))}(X^M)$ and (7.25). ∎

<u>Remark.</u> (7.25) states that for $\varepsilon, \alpha > 0$ and $f \in C$ there is a compact set K in E such that f and $U^\alpha f$ are continuous on K and

$$\sup_M P(\sup_t X^M_t(K^c) > \varepsilon) < \varepsilon.$$

This allows us to effectively treat f and $U^\alpha f$ as continuous functions and use the compact containment given by Proposition 7.7 to take limits in (7.4) (with $U^1 f$ in place of f, $f \in C$) and conclude that X^M converges weakly to Q_{m_0}, the unique solution of the martingale problem in Theorem 7.5.

<u>Theorem 7.13.</u> Assume Y is a Hunt process and the initial measures satisfy $(H_{7.2})$. Then $\{X^M : M \in \mathbb{N}\}$ converges weakly in $D(M_F(E))$ to Q_{m_0}, the law of the $(Y, -\sigma^2\lambda^2/2)$-superprocess starting at m_0 on $C(M_F(E))$. If $f \in \bigcup_{\alpha>0} U^\alpha(b\mathcal{E})$ and $\alpha > 0$, then

$$\langle X^M, U^\alpha f\rangle \xrightarrow{w} Q_{m_0}(\langle X, U^\alpha f\rangle \in .) \quad \text{in } D(\mathbb{R})$$

(where the limit is supported by $C(\mathbb{R})$).

<u>Proof.</u> Note that by $(H_{7.2})$,

$$(7.31) \quad P^{m_0^M}(Y \in A) = P^{m_0}(Y \in A)(K_M/(Mm_0(E))) \longrightarrow P^{m_0}(Y \in A).$$

$(H_{7.3})$ holds by Lemma 7.2. We claim $C = \bigcup_{\alpha>0} U^\alpha(b\mathcal{E})$ satisfies the remaining hypotheses of Theorem 7.12. Clearly $\bar{C}^{bp} = b\mathcal{E}$ and (7.24) is a simple consequence of $(H_{7.2})$ and the law of large numbers. (7.31) implies (7.23) with $m = cm_0$ and so Lemma 7.11 implies (7.25). Theorem 7.12 now gives the result. ∎

<u>Theorem 7.14.</u> Assume $\{P^{m_0^M}(Y \in .) : M \in \mathbb{N}\}$ is relatively compact in

$M_F(D(E))$, $(H_{7.1})$ holds, and there is a $\mathcal{C} \subset C_b(E)$ such that $\bar{\mathcal{C}}^{bp} = b\mathcal{E}$ and $U^\alpha:\mathcal{C}\longrightarrow C_b(E)$ for all $\alpha>0$. Then $\{X^M\}$ converges weakly in $D(M_F(E))$ to Q_{m_0}, the law of the $(Y,-\sigma^2\lambda^2/2)$-superprocess starting at m_0, on $C(M_F(E))$.

Proof. Again we only need to verify the hypotheses of Theorem 7.12. $(H_{7.3})$ holds by Lemma 7.2. The continuity of f and $U^\alpha f$ for f in $\mathcal{C}$, and Corollary 7.8 imply (7.25). Finally (7.24) holds by $(H_{7.1})$ and the continuity of $U^1 f$. ∎

We now use the above results to prove that

$H^M(t) = M^{-1}\sum_{\alpha\sim t}\delta_{(X^\alpha)^t_\cdot}$ converges weakly to the $(Y,-\sigma^2\lambda^2/2)$-historical process. Let $\hat{Y} = (D(\hat{E}),\hat{\mathcal{D}},\hat{\mathcal{D}}_{t+},\hat{Y}_t,\hat{P}^{(s,y)})$ be the càdlàg $\hat{E}$-valued Borel right process constructed from Y in Theorem 2.2.1 and Proposition 2.1.2. Recall from Section 2 that $\hat{W}_0:D(E)\longrightarrow D(\hat{E})$ is the continuous map $\hat{W}_0(y)(t) = (t,W_t(y)) = (t,y^t)$. If $x\in E$ let $w_x(t)\equiv x$ for all t and let $\hat{x} = \hat{W}_0(w_x)(0) = (0,w_x)\in \hat{E}$. If $m\in M_F(E)$ we embed E into $D(E)$ by the mapping $x \longrightarrow w_x$ and hence identify m with the image measure in $M_F(D(E))$. To avoid confusion we let $\hat{m} \in M_F(\hat{E})$ denote the image of m under the mapping $x \longrightarrow \hat{x}$. The definition of $\hat{Y}$ in Section 2 implies (see the Remark prior to Theorem 2.2.3)

$$(7.32)\qquad \hat{P}^{\hat{m}}(\hat{Y}\in A) = P^m(\hat{W}_0(Y)\in A).$$

Our main result is a simple application of Theorem 7.13 with $\hat{Y}$ in place of Y and $\hat{X}^M_t = M^{-1}\sum_{\alpha\sim t}\delta_{\hat{W}_0(Y^\alpha)(t)}1(X^\alpha_t\neq\Delta)$ in place of X^M_t.

Recall the definition of the (Y,Φ)-historical process following Theorem 2.2.3.

Theorem 7.15. Assume Y is a Hunt process and the initial measures satisfy $(H_{7.2})$. Then $\{H^M: M\in\mathbb{N}\}$ converges weakly in $D(M_F(D(E)))$ to Q_{0,m_0}, the law of the $(Y,-\sigma^2\lambda^2/2)$ historical process, starting at m_0 at time 0, on $C(M_F(D(E)))$.

Proof. We continue to work on $(\Omega,\mathcal{F},P_M)$. Let $\hat{\Omega}_1 = \hat{E}_\Delta^{\mathbb{N}}$, where $\hat{E}_\Delta = \hat{E}\cup\{\Delta\}$ and Δ is added as an isolated point. Let $\hat{P}_1^M$ be the image of P_1^M under the mapping $(x_i) \longrightarrow (\hat{x}_i)$ where $\hat{\Delta} = \Delta$. Since $\{P_1^M\}$ satisfies $(H_{7.2})$ for m_0, clearly $\{\hat{P}_1^M\}$ will satisfy $(H_{7.2})$ for $\hat{m}_0$. Let $\hat{Y}^\alpha = \hat{W}_0(Y^\alpha)\in D(\hat{E})$. (7.1) and (7.32) show that $P_M(((\hat{x}_i),(\hat{Y}^\alpha,N^\alpha)_{\alpha\in I})\in\cdot) = \hat{P}_M(\cdot)$, where $\hat{P}_M$ is the probability on $\hat{\Omega} = \hat{\Omega}_1\times(D(\hat{E})\times\mathbb{Z}_+)^I$ which is constructed in the same way as P_M but with $\hat{m}_0$, $\hat{P}_1^M$, $\hat{Y}$ and $\hat{E}$ in place of m_0, P_1^M, Y and E. Theorem 2.2.1(b) and Proposition 2.1.2 show that $\hat{Y}$ is a Hunt process because Y is. Apply Theorem 7.13 with $\hat{m}_0,\hat{X}^M$, and $\hat{Y}$ in place of m_0, X^M and Y, to conclude that $\{\hat{X}^M\}$ converges weakly to $\hat{Q}_{\hat{m}_0}$, the law of the $(\hat{Y},-\sigma^2\lambda^2/2)$-superprocess starting at $\hat{m}_0$. Recall that $\pi:\hat{E}\longrightarrow D(E)$ is the projection map and $\bar{\pi}:M_F(\hat{E}) \longrightarrow M_F(D(E))$ is the continuous map $\bar{\pi}(\nu) = \nu\circ\pi^{-1}$. If $\tilde{\pi}:D(M_F(\hat{E})) \longrightarrow D(M_F(D(E)))$ is given by $\tilde{\pi}(x)(t) = \bar{\pi}(x(t))$, then $\tilde{\pi}$ is continuous and so $H^M = \tilde{\pi}(\hat{X}^M)$ converges weakly in $D(M_F(D(E)))$ to $\hat{Q}_{\hat{m}_0}\circ\tilde{\pi}^{-1} = Q_{0,m_0}$ (see (2.1.11)). Theorem 2.2.3 (c) gives the continuity of the limit process. ∎

Theorem 7.16. (a) Assume

(F) $x \longrightarrow P^x(Y\in\cdot)$ is a continuous map from E to $M_F(D(E))$.

If $(H_{7.1})$ holds, then

(7.33) $\{H^M\}$ converges weakly in $D(M_F(D(E)))$ to Q_{0,m_0}, the law of the $(Y,-\sigma^2\lambda^2/2)$-historical process, starting at m_0 at time 0, on $C(M_F(D(E)))$,

(b) Conversely, if (7.33) holds whenever $m_0 \in M_F(E)$ and $(H_{7.1})$ holds, then (F) holds.

<u>Proof.</u> (a) Lemma 7.2 and (F) imply $P^{m_0^M} \longrightarrow P^{m_0}$ in $M_F(D(E))$ and so by (7.32)

$$\hat{P}^{\hat{m}_0^M}(\hat{Y}\in,) \longrightarrow \hat{P}^{\hat{m}_0}(\hat{Y}\in.) \text{ in } M_F(D(\hat{E})).$$

$(H_{7.1})$ implies that $\{\hat{P}_1^M\}$ satisfies $(H_{7.1})$ for $\hat{m}_0$. Let $\{\hat{U}^\alpha:\alpha>0\}$ denote the resolvent of $\hat{Y}$. We will show that for $\alpha>0$, $\hat{U}^\alpha:C_b(\hat{E}) \longrightarrow C_b(\hat{E})$. The result then follows as for Theorem 7.15 only now one uses Theorem 7.14 with $C = C_b(\hat{E})$ instead of Theorem 7.13.

Recall (see the Remark prior to Theorem 2.2.3) that $\hat{Y}$ has semigroup

$$\hat{P}_t g(s,y) = P^{y(s)}(g(s+t,y/s/Y^t)), \quad (s,y)\in \hat{E}, \ g\in b\hat{\mathcal{E}}.$$

Fix $g\in C_b(\hat{E})$ and $t\geq 0$. Let $F = \{(s,y,w)\in\hat{E}\times D(E):w(0)=y(s)\}$ and define $h:F\longrightarrow \mathbb{R}$ by $h(s,y,w) = g(s+t,y/s/w^t)$. It is easy to see that if $w(t) = w(t-)$, then $(s,y,w)\in F$ is a continuity point of h. Let $(s_n,y_n) \longrightarrow (s,y)$ in $\hat{E}$ and fix t such that $Y(t-)=Y(t)$, $P^{y(s)}$-a.s. Since y_n (respectively, y) is constant on $[s_n,\infty)$ (respectively, on $[s,\infty)$) and $y_n \to y$ in D(E), we have $y_n(s_n) \to y(s)$ and so $P^{y_n(s_n)}(Y\in.) \longrightarrow P^{y(s)}(Y\in.)$ in $M_F(D(E))$ by (F). By Skorokhod's theorem we may choose $\{Y_n\}$, Y_∞ such that Y_n has law $P^{y_n(s_n)}(Y\in.)$, Y_∞ has law $P^{y(s)}(Y\in.)$ and $Y_n \xrightarrow{a.s.} Y_\infty$. Then

$$\hat{P}_t g(s_n,y_n) = P^{y_n(s_n)}(h(s_n,y_n,Y))$$

$$= P(h(s_n,y_n,Y_n)) \longrightarrow P(h(s,y,Y_\infty)) = \hat{P}_t g(s,y)$$

by the above continuity result because (s_n,y_n,Y_n) and $(s,y,Y) \in F$

and $Y(t)=Y(t-)$ a.s. As there is at most a countable exceptional set of times, we have shown $\hat{U}^{\alpha}g(s_n,y_n) \longrightarrow \hat{U}^{\alpha}g(s,y)$ and the proof of (a) is complete.

(b) Let $x_M \to x$ in E, and let $P_1^M = \delta_{(x_i^M)}$ where $x_i^M=x_M$ $i\le M$ and $x_i^M = \Delta$ if $i>M$. Then $X_0^M = \delta_{x_M}$ and $(H_{7.1})$ holds for $m_0 = \delta_x$. If $f\in C_b(D(E))$ then $\langle H^M,f\rangle \xrightarrow{w} \langle H,f\rangle$ on $D(\mathbb{R})$ where H has law Q_{0,δ_x} (by (7.33)). Lemma 7.2 allows us to takes means and conclude that $\lim_{M\to\infty} P_M(\langle H_t^M,f\rangle) = Q_{0,\delta_x}(\langle H_t,f\rangle)$ for all $t\ge 0$. By Lemma 7.3(b), Theorem 2.1.5(d) and (2.2.1) this means that

$$\lim_{M\to\infty} P^{x_M}(f(Y^t)) = P^x(f(Y^t)) \quad \text{for all } t\ge 0.$$

Standard arguments (e.g. Ethier-Kurtz (1986, Ch. 3, Thm. 7.2,7.8)) now imply (F).■

Recall Y is a Feller process if E is locally compact and its semigroup is strongly continuous on the continuous functions which vanish at ∞.

Corollary 7.17. Assume Y is a Feller process and $(H_{7.1})$ holds. Then (7.33) holds.

Proof. (F) holds by Ethier-Kurtz (1986, Ch.4, Thm. 2.5).■

Remark 7.18. The results of this section remain valid if E is a metrizable Lusin space (i.e. homeomorphic to a Borel subset of a compact metric space). This was the setting in which Fitzsimmons (1988,1989) proved existence and uniqueness for the martingale problem in Theorem 7.5. D(E) is then a cosouslin metrizable space (homeomorphic to the complement of an analytic subset of a compact metric space). Probabilities on D(E) and, more generally, weakly convergent sequences of probabilities on D(E) remain tight (see Dellacherie-Meyer (1978,III.59) and the references cited there). These observations are needed to extend tightness results such as Corollary 7.8 and Lemma 7.10 to this setting.

We close this section with a nonstandard formulation of the weak convergence result Theorem 7.15. Although this result is immediate from Theorem 7.15 and the nonstandard characterization of weak convergence, in fact it was essentially proved first in our derivation of Theorem 7.15 (see the proof of 7.12).

Fix $M \in {}^*\mathbb{N}-\mathbb{N}$ and let $({}^*\Omega,\mathcal{G},P)$ denote the Loeb space $({}^*\Omega,L({}^*\mathcal{F}),L({}^*P_M))$.

<u>Theorem</u> <u>7.19.</u> Assume Y is a Hunt process and $(H_{7.2})$. There is a unique continuous $M_F(D(E))$-valued process H on $({}^*\Omega,\mathcal{G},P)$ such that

(7.34) $$H_{°t}(A) = L(H_t^M)(st^{-1}_{D(E)}(A)) \text{ for all } A \text{ in } \mathcal{B}(D(E)), \text{ and}$$
$$L(H_t^M)(ns({}^*D(E))^c) = 0, \text{ for all } °t < \infty \text{ a.s.}$$

Moreover, the law of H on $C(M_F(D(E)))$ is Q_{0,m_0}, the law of the $(Y,-\sigma^2\lambda^2/2)$-historical process starting at m_0 at time zero.

8. A Modulus of Continuity for the Supports of Super-diffusions.

If $A \subset E$, let $A^{\delta} = \{x \in E : d(x,A) \le \delta\}$. The following one-sided modulus of continuity for the supports of super-Brownian motion was derived in [DIP, Theorem 1.1] (Q_m is the law of super-Brownian motion):

(8.0) If $h(u) = (u\log^{+}1/u)^{1/2}$ and $c>2$, then Q_m-a.s. there is a $\delta(c,\omega) > 0$ such that if $0 < t-s < \delta(c,\omega)$ then $S(X_t) \subset S(X_s)^{ch(t-s)}$.

This result followed from a stronger uniform modulus of continuity (for which $c>2$ is sharp) on the paths of the branching particle systems discussed in Section 7. The historical process will allow us to formulate a continuity result which reflects this stronger modulus of continuity and for which $c>2$ is sharp (see Theorem 8.7), something that was impossible to do in terms of X alone. This continuity result played an important part in the proofs of various path properties of super-Brownian motion in [DIP] and Perkins (1989,1990). We will take this opportunity to extend this result to a wide class of super-diffusions and establish some of the simpler path properties for this class and their associated historical processes.

We first work with the approximating processes X^M and H^M defined on $(\Omega,\mathcal{F},P_M)$ in Section 7. Dependence on M is sometimes suppressed (e.g. P denotes P_M). Those familiar with nonstandard analysis should take M to be infinite and work on the obvious Loeb space.

Throughout this section $\sigma^2>0$ and Y denotes an E-valued diffusion, i.e. a Borel strong Markov process with continuous sample paths taking values in the Polish space E. Let $\rho(\delta,t) = \sup_{x\in E} P^x(d(Y_t,Y_0)\geq\delta)$ and

$\mathcal{H} = \{h:[0,1]\to[0,\infty): h$ continuous, non-decreasing and $h(0) = 0\}$.

<u>Theorem 8.1.</u> Assume $(H_{7.2})$, Y is a diffusion and let $h\in\mathcal{H}$ satisfy

$$(8.1)\quad \sum_{n=1}^{\infty} \rho(h(2^{-n}),2^{-n})4^n < \infty$$

$$(8.2)\quad \int_0^1 h(r)(\log 1/r)r^{-1}dr < \infty.$$

If $G(u) = (2/\log 2)^2\int_0^1 h(2ur)(\log 1/r)r^{-1}dr$, then G is continuous on $[0,1/2)$ and P_M-a.s.

(8.3) there is a $\delta_M(\omega) > 0$ such that if $0 < t-s < \delta_M(\omega)$, $\beta\sim t$ and $X_t^\beta \neq \Delta$, then $d(X_t^\beta,X_s^\beta) \leq G(t-s)$.

Also if $\varepsilon_n = \sum_{j\geq n} \rho(h(2^{-j}),2^{-j})4^j$, then $\lim \varepsilon_n = 0$ and δ_M in (8.3) satisfies

$$(8.4)\quad P_M(\delta_M\leq 2^{-n}) \leq c_{8.0}(\sigma^2)m_0^M(E)\varepsilon_n^{1/2}$$

($c_{8.0}(\sigma^2)$ is independent of M).

<u>Remarks 8.2.</u> (a) It is of course critical that the upper bound in (8.4) remains bounded as $M\to\infty$ and hence will allow us to let $M\to\infty$ in (8.3).

(b) Under slightly stronger hypotheses, a sharper result (which is best possible when Y is Brownian motion) will be derived from this theorem in Theorem 8.4 below.

(c) If $\sum_{n=1}^{\infty} \rho(n^{-2-\delta},2^{-n})4^n < \infty$ for some $\delta > 0$,

then (8.1) and (8.2) hold with $h(u) = \left(\log(\frac{1}{u} \vee e)/\log 2\right)^{-2-\delta}$

and $G(u) \le c(\delta)(\log 1/u)^{-\delta}$.

(d) It is easy to show (see the proof of Theorem 8.1) that (8.1) and (8.2) imply that a right continuous Borel strong Markov process has continuous paths a.s. and hence the condition that Y is a diffusion is redundant.

Notation. If $t>0$, $\theta\in(0,1)$, $u\in(0,1]$ and $h\in\mathcal{H}$, let

$$B_n(t,u,\theta,h) = B_n(t) = \{\omega: d(X^\beta_{t-\theta^k u}, X^\beta_{t-\theta^{k-1}u}) \ge h(\theta^k u(\theta^{-1}-1)) \text{ for some } \beta\sim t,\ X^\beta_t\ne\Delta,\ k>n \text{ and } \theta^{k-1}u\le t\}.$$

The proof of the following estimate is the same as the derivation of Lemma 4.3 in [DIP] (the modifications needed to handle a general $\sigma^2>0$ are trivial.)

Lemma 8.3. (a) $P(B_n(t,u,\theta,h))$

$$\le c_{8.1}(u,\theta)m_0^M(E)\sum_{k>n}\theta^{-k}\rho(h(\theta^k u(\theta^{-1}-1)),\theta^k u(\theta^{-1}-1)).$$

(b) If $\omega\notin B_n(t,u,\theta,h)$ and $\theta^n u\le t$, then

$$d(X^\beta_t, X^\beta_{t-\theta^n u}) \le \sum_{k=n+1}^{\infty} h(\theta^k u(\theta^{-1}-1)).$$

Proof of Theorem 8.1. Let $B_n(t) = B_n(t,1,1/2,h)$, and for $N\in\mathbb{N}$ let

$$C_n(N) = \bigcup_{k=n}^{\infty}\bigcup_{i=1}^{N2^k} B_k(i2^{-k}),\quad K(N) = \min\{n: \omega\notin C_n(N)\}\ (\min\emptyset=\infty).$$

Then

$$\text{(8.5)}\quad P(K(N)\ge n) \le P(C_{n-1})$$

$$\le c_{8.1}m_0^M(E)\sum_{k=n-1}^{\infty} N2^k\sum_{j>k}2^j\rho(h(2^{-j}),2^{-j})$$

$$\le c_{8.1}m_0^M(E)N\varepsilon_n,$$

where ε_n (as in the statement of the theorem) decreases to zero by (8.1). Therefore $K(N) < \infty$ a.s. and we may fix ω outside a P-null set such that $K(N) < \infty$ for all $N \in \mathbb{N}$. Let $s,t \in [0,N]$ ($N \in \mathbb{N}$ fixed) and choose $n \geq K(N)$ such that

$$2^{-n-1} < t-s \leq 2^{-n}.$$

A well-known argument of Lévy using the binary expansion of t and s (see the proof of Proposition 4.4 of [DIP]) now shows that the choice of ω and Lemma 8.3(b) implies that if $\beta \sim t$ and $X_t^\beta \neq \Delta$, then

$$d(X_t^\beta, X_s^\beta) \leq \sum_{k=n+1}^{\infty} h(2^{-k}) + 2 \sum_{k=n+1}^{\infty} \sum_{j>k} h(2^{-j})$$

$$\leq 2 \sum_{j>n} (j-n) h(2^{-j}).$$

An elementary argument shows that for $j>n$,

$$(j-n) \leq 2(\log 2)^{-2} \int_{2^{-j}}^{2^{1-j}} t^{-1} \log(1/2^n t)\,dt$$

and so

$$d(X_t^\beta, X_s^\beta) \leq 4(\log 2)^{-2} \sum_{j>n} \int_{2^{-j}}^{2^{1-j}} h(t) t^{-1} \log(1/2^n t)\,dt$$

$$= 4(\log 2)^{-2} \int_0^{2^{-n}} h(t) t^{-1} \log(1/2^n t)\,dt$$

$$= 4(\log 2)^{-2} \int_0^1 h(2^{-n}u) u^{-1} \log(1/u)\,du$$

(8.6) $$\leq G(t-s).$$

Let $N_0(\omega) = \min\{n : X_n^M(\omega) = 0\}$ and $\delta(\omega) = 2^{-K(N_0(\omega))} > 0$ a.s. The above argument proves (8.3). Also

$$P(\delta \leq 2^{-n}) \leq P(N_0 \geq N) + P(K(N) \geq n)$$

$$\leq P(X_{N-1}^M \neq 0) + c_{8.1} m_0^M(E) N \varepsilon_n \quad \text{(by (8.5))}$$

$$\le c(\sigma^2)m_0^M(E)(N-1)^{-1} + c_{8.1}m_0^M(E)N\varepsilon_n$$

(Harris (1963, p. 21-22)).

Take $N\in[\varepsilon_n^{-1/2}+1,\varepsilon_n^{-1/2}+2]\cap\mathbb{N}$ to complete the proof of (8.4). ∎

<u>Theorem 8.4.</u> Assume $(H_{7.2})$, Y is a diffusion and $h\in\mathcal{H}$ satisfies

(8.7) $\sum_{j=1}^{\infty}\theta^{-2j}\rho(h(\theta^j r),\theta^j r) < \infty$ for all $\theta\in(0,1)$ and $r>0$,

(8.8) if $f(r) = \sup\{h(ur)/h(u):0<u\le 1, h(u)>0\}$ for $0<r\le 1$, then $\int_0^1 f(r)(\log 1/r)r^{-1}dr < \infty$.

Then P_M-a.s. for any $c>1$ there is a $\delta_M(c,\omega) > 0$ such that

(8.9) if $0 < t-s \le \delta_M(c,\omega)$, $\beta\sim t$, and $X_t^\beta\ne\Delta$, then $d(X_t^\beta,X_s^\beta) < ch(t-s)$,

(8.10) there is a function $p:(1,\infty)\times[0,\infty)\longrightarrow[0,1]$ independent of the choice of M such that

$$P_M(\delta_M(c)\le\rho) \le m_0^M(E)p(c,\rho) \quad\text{and}\quad \lim_{\rho\downarrow 0}p(c,\rho) = 0.$$

<u>Proof.</u> Let $c_1 = \int_0^1 f(w)w^{-1}dw$, $c_2 = \int_0^1 f(w)(\log 1/w)w^{-1}dw$, $c>1$,

choose $\theta\in(0,1/2]$ $(\theta=\theta(c))$ sufficiently small so that

(8.11) $f(\theta)(1+c_1(1-\theta)^{-1}) < (c-1)/2$

and then choose $N_1\in\mathbb{N}$ $(N_1=N_1(c))$ sufficiently large so that

(8.12) $2/N_1 < \theta$

and

(8.13) $8c_2(\log 2)^{-2}f(2/\theta N_1) < (c-1)/2.$

If $N\in\mathbb{N}$ let

$$D_n=D_n(c,N) = \bigcup_{k=n}^{\infty}\bigcup_{i=0}^{[\theta^{-k}N]}\bigcup_{p=0}^{N_1-1}\bigcup_{q=1}^{N_1} B_k((i+(p/N_1))\theta^k,q/N_1,\theta,h),$$

$$J(\omega,N) = \min\{n:\omega\notin D_n\}$$

and

$$c_{8.2}(c) = \max\{c_{8.1}(q/N_1,\theta):1\le q\le N_1\}.$$

Then Lemma 8.3(a) shows that

$$P(J\ge n) \le P(D_{n-1})$$

$$\le \sum_{q=1}^{N_1} N_1 c_{8.2} m_0^M(E) \sum_{k=n}^{\infty} (\theta^{-k}N+1)$$

$$\cdot \sum_{j>k} \theta^{-j}\rho(h(\theta^j(q/N_1)(\theta^{-1}-1)),\theta^j(q/N_1)(\theta^{-1}-1))$$

$$\le 2c_{8.2}NN_1 m_0^M(E) \sum_{q=1}^{N_1} \sum_{j=n}^{\infty} \theta^{-2j}\rho(h(\theta^j(q/N_1)(\theta^{-1}-1)),\theta^j(q/N_1)(\theta^{-1}-1))$$

$$= Nm_0^M(E)\varepsilon_n(c)$$

where $\varepsilon_n(c)\downarrow 0$ as $n\to\infty$ by (8.7). If $K(\omega,N)$ is as in (8.5) and $\delta_1(\omega,N) = \min(\theta^{J(N)},2^{-K(N)})$, then, by the above and (8.5),

(8.14) $P(\delta_1\le\theta^n) \le P(K(N)\ge n) + P(J(N)\ge n) \le m_0^M(E)N(c_{8.1}\varepsilon_n+\varepsilon_n(c))$

where ε_n are as in the statement of Theorem 8.1.

Assume $s,t\in[0,N]$ satisfy $0< t-s < \delta_1(\omega,N)$, let $\beta\sim t$ satisfy $X_t^\beta(\omega)\ne\Delta$ and choose $n\ge J(\omega,N)$ such that

(8.15) $\theta^{n+1} < t-s \le \theta^n$.

Choose $0\le i\le\theta^{-n}N$ and $p\in\{0,1,\dots,N_1-1\}$ such that

$$t_0 = (i+(p/N_1))\theta^n \le t < (i+((p+1)/N_1))\theta^n.$$

By (8.12) and (8.15) there is a $q\in\{1,\dots,N_1\}$ such that

$$(i+(p-q-1)/N_1)\theta^n < s \le (i+(p-q)/N_1)\theta^n = s_0 < t_0.$$

Since $n\ge J(\omega,N)$ we have $\omega\notin B_n(t_0,q/N_1,\theta,h)$ and therefore

$$d(X_{t_0}^\beta,X_{s_0}^\beta) = d(X_{t_0}^\beta,X_{t_0-(q/N_1)\theta^n}^\beta)$$

$$\le \sum_{k=n+1}^{\infty} h(\theta^k(q/N_1)(\theta^{-1}-1)) \qquad \text{(by Lemma 8.3(b))}$$

$$\le h(t-s) + h(\theta(t-s))$$

$$+ \sum_{k=n+2}^{\infty} \int 1(\theta^k q/N_1 \le v \le \theta^{k-1} q/N_1) h(v) [\theta^{k-1}(q/N_1)(1-\theta)]^{-1} dv$$

$$(8.16) \quad \le h(t-s) + h(\theta(t-s)) + (1-\theta)^{-1} \int 1(0 \le v \le \theta^{n+1} q/N_1) h(v) v^{-1} dv.$$

Since $t-t_0 \le t-s \le 2^{-K(N)}$, (8.6) implies

$$d(X^\beta_t, X^\beta_{t_0}) \le G(t-t_0) \le G(\theta^n/N_1).$$

This and the same bound for $d(X^\beta_s, X^\beta_{s_0})$, combine with (8.16) to show

$$d(X^\beta_t, X^\beta_s) \le h(t-s) + h(\theta(t-s)) + (1-\theta)^{-1} \int_0^1 h(\theta^{n+1}(q/N_1)w) w^{-1} dw$$

$$+ 8(\log 2)^{-2} \int_0^1 h(2\theta^n w/N_1)(\log 1/w) w^{-1} dw$$

$$\le h(t-s) + f(\theta)h(t-s)$$

$$+ (1-\theta)^{-1} h(\theta^{n+1} q/N_1) c_1 + 8(\log 2)^{-2} h(2\theta^n/N_1) c_2$$

$$\le h(t-s) + f(\theta)h(t-s)$$

$$+ (1-\theta)^{-1} f(\theta) h(t-s) c_1 + 8(\log 2)^{-2} f(2/(\theta N_1)) h(t-s) c_2$$

$$(8.17) \quad \le ch(t-s)$$

by (8.11) and (8.13). To remove the restriction $s,t \in [0,N]$ set $N_0(\omega) = \min\{n: X^M_n(\omega)=0\}$ and $\delta(c,\omega) = \delta_1(\omega, N_0(\omega))$. (8.9) is then immediate from (8.17). (8.10) follows easily from (8.14) just as in the end of the proof of Theorem 8.1. ∎

<u>Remarks</u> <u>8.5.</u> (a) If $\rho(h(t),t)t^{-2}$ is decreasing near zero then (8.7) (and (8.1)) is equivalent to $\int_0^{t_0} \rho(h(t),t)t^{-3} dt < \infty$ for some $t_0>0$. More generally if $\rho \le \hat{\rho}$, $\hat{\rho}(h(t),t)t^{-2}$ is decreasing near zero and $\int_0^{t} \hat{\rho}(h(t),t)t^{-3} dt < \infty$ for some $t_0>0$, then (8.7) holds.

(b) (8.8) clearly implies (8.2) since $h(r) \le f(r)h(1)$. The slight strengthening gives a result which is best possible when Y is a Brownian motion in $\mathbb{R}^d$. In this case (8.7) holds with $h(t) = c(t((\log 1/t)\vee 1))^{1/2}$ and $c>2$. Hence the Theorem implies Theorem 4.5 of [DIP] which states that for any $c>2$ there is a $\delta_M(c,\omega)>0$ such that

$$(8.18)\quad 0 < t-s < \delta_M(c,\omega),\ \beta\sim t \text{ and } X_t^\beta \ne \Delta$$
$$\Longrightarrow |X_t^\beta - X_s^\beta| \le c((t-s)\log^+ 1/(t-s))^{1/2}$$
$$\lim_{\rho\downarrow 0}\sup_M P(\delta_M(c,\omega)\le\rho) = 0.$$

This result is shown to be false if $c<2$ in Theorem 4.6 of [DIP]. It is perhaps worth noting that Theorem 8.1 already gives (8.18) except for the exact value of the constant c.

We next interpret Theorems 8.1 and 8.4 in terms of the historical process. We first use the nonstandard model for the historical process (Theorem 7.19, but it is easy to use weak convergence arguments) to transfer Theorems 8.1 and 8.4 to the limit. We consider only Theorem 8.4, the analogue of Theorem 8.1 is then obvious.

Throughout the rest of this section H_t is the $(Y,-\sigma^2\lambda^2/2)$-historical process (with law Q_{0,m_0}), and $X_t = \bar{\Pi}_t(H_t)$ is the ordinary $(Y,-\sigma^2\lambda^2/2)$-superprocess. Since Y is a diffusion H_t is a finite random measure on $C(E)$.

<u>Notation.</u> If $h\in\mathcal{H}$, $\delta\in[0,1]$ and $T\in[0,\infty]$ let

$$K(\delta,h,T) = \{y\in C(E): d(y(t),y(s)) \le h(t-s) \text{ for all } s,t\in[0,T] \text{ such that } 0<t-s\le\delta\},$$

and

$$K(\delta,h) = K(\delta,h,\infty).$$

Theorem 8.6. Assume the hypotheses of Theorem 8.4. If $c>1$ then Q_{0,m_0}-a.s. there is a $\delta(c,\omega)$ such that $S(H_t) \subset K(\delta(c),ch)$ for all $t>0$.

Proof. We work in the nonstandard setting of Theorem 7.19. Let $c>1$, $\delta_M(c)$ be as in Theorem 8.4 (but with M in $^*\mathbb{N}-\mathbb{N}$) and let $\delta(c) = {}^\circ\delta_M(c) > 0$ P-a.s. by (8.10). $K(\delta(c),ch)$ is clearly closed. Fix ω such that (7.34) holds and $\delta(c)>0$. If $t>0$, then

$$H_t(K(\delta(c),ch)^c) = L(H_t^M)(st^{-1}(K(\delta(c),ch)^c)) \le {}^\circ H_t^M({}^*K(\delta(c),ch)^c) \quad \text{(K is closed)}$$

$$= 0,$$

the last by (8.9) and the Transfer Principle. ∎

Theorem 8.7. Let Q_{0,m_0} be the law of the $(Y,-\sigma^2\lambda^2/2)$-historical process, H_t, where $H_0 = m_0$ and Y is a Brownian motion on $\mathbb{R}^d$. Let $h_0(u) = (u((\log 1/u)\vee 1))^{1/2}$.

(a) If $c>2$, then Q_{0,m_0}-a.s. there is a $\delta(c,\omega)$ such that $S(H_t) \subset K(\delta(c),ch_0)$ for all $t>0$.

(b) If $c<2$, then Q_{0,m_0}-a.s. for all $\delta>0$ there is a t in $[0,1]$ such that $H_t(K(\delta,ch_0,1)^c) > 0$.

Proof. (a) is immediate from Theorem 8.6 (see Remark 8.5(b)).

For (b) we use the following lemma, whose proof is an easy modification of that of Theorem 4.6 of [DIP]. It deals with the nonstandard model in Theorem 7.19.

Lemma 8.8. Let $\eta \in {}^*\mathbb{N}-\mathbb{N}$ and $M = 2^\eta$. Assume $(H_{7.2})$ and Y is a Brownian motion on $\mathbb{R}^d$. If $c_0<2$, then $L({}^*P_M)$-a.s. there is a $N(\omega) \in \mathbb{N}$ such that if $n \ge N(\omega)$, then

$$\sup\{|X^\beta(2j2^{-n})-X^\beta((2j-1)2^{-n})|h_0(2^{-n})^{-1}: (2j+1)2^{-n} \le 1,\ \beta\sim 2j2^{-n},$$

$$M^{-1} \sum_{\substack{\gamma \sim (2j+1)2^{-n} \\ \gamma > \beta}} 1(X^{\gamma}((2j+1)2^{-n}) \neq \Delta) > 2^{-n}\}$$

$> c_0.$

Proof of (b). We work in the nonstandard setting of Theorem 7.19 where $M = 2^{\eta}$ as in the above lemma. Let $c \in (0,2)$ and choose $c_0 \in (c,2)$. Fix ω outside a null set such that (7.34) and the conclusion of Lemma 8.8 hold. If $\delta > 0$ choose $n \geq N(\omega)$ such that $2^{-n} < \delta$ and then $j \in \mathbb{N}$ with $(2j+1)2^{-n} \leq 1$ and $\beta \sim 2j2^{-n}$ such that

(8.19) $$|X^{\beta}(2j2^{-n}) - X^{\beta}((2j-1)2^{-n})| > c_0 h_0(2^{-n})$$

(8.20 $$M^{-1} \sum_{\substack{\gamma \sim (2j+1)2^{-n} \\ \gamma > \beta}} 1(X^{\gamma}((2j+1)2^{-n}) \neq \Delta) = Z^{n,\beta} > 2^{-n}.$$

Let

$$K_n = \{y \in C(\mathbb{R}^d) : \max_{1 \leq j \leq 2^n} |y(j2^{-n}) - y((j-1)2^{-n})| \leq c_0 h_0(2^{-n})\}.$$

(8.19) implies

(8.21) $$H^M((2j+1)2^{-n})({}^*K_n^c) \geq Z^{n,\beta} > 2^{-n}.$$

Therefore

$$H((2j+1)2^{-n})(K(\delta, ch_0, 1)^c) = L(H^M((2j+1)2^{-n}))(st^{-1}_{C(\mathbb{R}^d)}(K(\delta, ch_0, 1)^c))$$

(by (7.34))

$$\geq L(H^M((2j+1)2^{-n}))({}^*K_n^c)$$

$$\geq 2^{-n} \quad \text{(by (8.21))}. \blacksquare$$

Proposition 8.9. Assume the hypotheses of Theorem 8.1 and let $M = 2^{\eta}$ where $\eta \in {}^*\mathbb{N} - \mathbb{N}$. Then $L({}^*P_M)$-a.s. for any $0 < {}^{\circ}t < \infty$ and $\gamma \sim t$, $X_t^{\gamma} \neq \Delta$ implies $X^{\gamma}_{\cdot \wedge t}$ is an S-continuous mapping from ${}^*[0,\infty)$ to ${}^*E$.

Proof. If $0 < \varepsilon \leq t$ let

$$I(t,\varepsilon) = \{\gamma \in I : \gamma \sim t-\varepsilon, \ \exists \beta \sim t \text{ such that } \gamma < \beta \text{ and } X_t^{\beta} \neq \Delta\}$$

and let $Z(t,\varepsilon)$ denote the cardinality of $I(t,\varepsilon)$. Let $p>3$ and if $\gamma\in I$, $\varepsilon>0$ let

$$Z^{\gamma}(\varepsilon) = M^{-1} \sum_{\substack{\beta\sim(|\gamma|/M)+\varepsilon \\ \beta>\gamma}} \mathbf{1}(X^{\beta}(|\gamma|/M + \varepsilon)\neq\Delta).$$

Then an elementary argument (see for example [DIP Lemma 4.8]) shows that (we write P for $L({}^{*}P_M)$)

$$P(Z^{\gamma}(2^{-n}) \le 2^{-np} \text{ for some } \gamma\in I(j2^{-n},2^{-n}) \text{ and } 1\le j\le n2^{-n}) \le cn2^{-n(p-3)}$$

and hence by Borel-Cantelli

(8.22) P-a.s. there is an $N(\omega)\in\mathbb{N}$ such that for all $n\ge N$, $1\le j\le n2^{-n}$ and $\gamma\in I(j2^{-n},2^{-n})$, $Z^{\gamma}(2^{-n}) > 2^{-np}$.

Fix ω outside a P-null set such that (8.22), (7.34) and (8.3) hold with ${}^{\circ}\delta_M > 0$ (the latter by the Transfer Principle and (8.4)). Let $0< {}^{\circ}t < \infty$, $\gamma \sim t$ and $X_t^{\gamma}\neq\Delta$. Let $0\le s<t$ with ${}^{\circ}s < {}^{\circ}t$ and choose n in $\mathbb{N}$ sufficiently large and $j\in\mathbb{N}$ such that $s \le v = (j-1)2^{-n} < j2^{-n} = u < t \le n$ and $n\ge N(\omega)$. If $\beta=\gamma|Mv$, then $\beta\in I(u,2^{-n})$ and

$$H_u^M(\{y:y^s=(X^{\gamma})^s\}) \ge H_u^M(\{y:y^v=(X^{\beta})^v\}) \ge Z^{\beta}(2^{-n}) > 2^{-np}$$

(by (8.22)).

Since $L(H_u^M)$ is supported by $ns({}^{*}D(E))$ (by (7.34)) this implies that $(X^{\gamma})^s\in ns({}^{*}D(E))$ for all ${}^{\circ}s<{}^{\circ}t$. (8.3) shows that $(X^{\gamma})^s$ is S-continuous for all ${}^{\circ}s<{}^{\circ}t$. Let ${}^{\circ}s_n\uparrow {}^{\circ}t$ (${}^{\circ}s_n < {}^{\circ}t$). Then (8.3) implies that $d(X_{s_n}^{\gamma},X_t^{\gamma}) \le G(t-s_n)$ for n sufficiently large. Therefore $st_E(X_{s_n}^{\gamma})$ is a Cauchy sequence in E (note that the S-continuity of $(X^{\gamma})^{s_n}$ is used to show $X_{s_n}^{\gamma} \in ns({}^{*}E)$) and hence converges to some $x\in E$. Since ${}^{\circ}d(x,X_t^{\gamma}) \le {}^{\circ}d(x,X_{s_n}^{\gamma}) + {}^{\circ}d(X_{s_n}^{\gamma},X_t^{\gamma})$ which approaches 0 as $n\to\infty$, $X_t^{\gamma}\in ns({}^{*}E)$ and $st_E(X_t^{\gamma}) = x$. By

(8.3) we see that $X^{\gamma}_{t'} \in ns(*E)$ and $st_E(X^{\gamma}_{t'}) = st_E(X^{\gamma}_t)$ for all $t' \approx t$, $t' \le t$. The proof of S-continuity is complete. ∎

<u>Theorem 8.10.</u> Assume the hypotheses of Theorem 8.1. Then Q_{0,m_0}-a.s.

(a) $S(H_t)$ is compact in $C(E)$ for all $t>0$,

(b) $S(X_t) = \Pi_t(S(H_t))$ for all $t \ge 0$,

(c) $S(X_t)$ is compact in E for all $t>0$.

<u>Proof.</u> (a) We work in the nonstandard setting of Theorem 7.19. Fix ω outside a null set such that (7.34) and the conclusion of Proposition 8.9 are both true. If $t \in (0,\infty)$ then $S(H^M_t) = \{(X^{\gamma})^t : \gamma \sim t, X^{\gamma}_t \ne \Delta\}$ is an internal subset of $ns({}^*C(E))$ and so (see Luxemburg (1969, Thm 3.6.1)) $st_{C(E)}(S(H^M_t))$ is compact in $C(E)$. (7.34) implies $S(H_t) \subset st_{C(E)}(S(H^M_t))$ and (a) follows.

(b) Since $X_t = H_t \circ \Pi_t^{-1}$ the inclusion $\Pi_t(S(H_t)) \subset S(X_t)$ is immediate, as is the result for $t=0$. It is obvious that $X_t(\Pi_t(S(H_t))^c) = 0$. Since $\Pi_t(S(H_t))$ is compact in E for all $t>0$ a.s. by (a) both (b) and (c) follow. ∎

The following result is a standard reformulation of a nonstandard result in [DIP Lemma 4.9]. The proof given there for Y a d-dimensional Brownian motion extends without difficulty to our more general setting thanks to Proposition 8.9. The condition $0 < {}^\circ s$ in Lemma 4.9 of [DIP] is readily removed.

<u>Proposition 8.11.</u> Assume the hypotheses of Theorem 8.1. Q_{0,m_0}-a.s. for any $0 \le s \le t$ and any $y \in S(H_t)$, $y^s \in S(H_s)$ and $y_s \in S(X_s)$.

We now are ready to extend (8.0) both to the more general superprocesses of Theorems 8.1 and 8.4 and to their historical processes. We consider the more precise Theorem 8.4.

Convention. If $A \subset C(E)$ and $\delta>0$ we modify our earlier definition and let

$$A^\delta = \{y \in C(E): \sup_t d(y(t),w(t)) \le \delta \text{ for some } w \in A\}.$$

Theorem 8.12. Assume the hypotheses of Theorem 8.4. Q_{0,m_0}-a.s. if $c > 1$ there is a $\delta(c,\omega)$ such that if $s,t \ge 0$ satisfy $0 \le t-s < \delta(c,\omega)$, then

$$S(H_t) \subset S(H_s)^{ch(t-s)}$$

and

$$S(X_t) \subset S(X_s)^{ch(t-s)}.$$

Proof. It clearly suffices to consider a fixed $c>1$. Fix ω outside a null set so that the conclusions of Theorem 8.6 and Proposition 8.11 both hold. Assume $0 \le t-s < \delta(c,\omega)$ and $y \in S(H_t)$, Then Theorem 8.6 implies (recall $H_t \in M_F(C(E))^t$) $\sup_u d(y^s_u, y_u) \le ch(t-s)$. Since $y^s \in S(H_s)$ by Proposition 8.11 we have proved $S(H_t) \subset S(H_s)^{ch(t-s)}$. The second conclusion is now immediate from this and Theorem 8.10(b). ∎

If (S,ρ) is a metric space let $(\mathcal{K}(S),\bar{\rho})$ be the space of compact subsets of S with the Hausdorff metric $\bar{\rho}$ induced by ρ (see Dugundji (1966, p.209), Cutler (1984), Perkins (1990, Section 1)). We let $S^H_t = S(H_t)$, $S^X_t = S(X_t)$.

Theorem 8.13. Assume the hypotheses of Theorem 8.1. Q_{0,m_0}-a.s.

(a) $\{S(H_t):t>0\}$ has right continuous paths with left limits in $\mathcal{K}(C(E))$.

(b) $\{S(X_t):t>0\}$ has right continuous paths with left limits in $\mathcal{K}(E)$.

(c) $S^H_t \subset S^H_{t-}$ and $S^X_t \subset S^X_{t-}$ for all $t>0$.

Proof. This is proved for super-Brownian motion in Perkins (1990,

Theorem 1.4) (see also [DIP, Theorem 1.2]). The proof given there goes through with one minor change in the proof of the existence of left-hand limits. In Lemma 4.1 of Perkins (1990) we used the fact that closed balls B in $\mathbb{R}^d$ are compact. Fix $\delta>0$ and use Proposition 8.9 to see that

$$st_{C(E)}(\{N^{\gamma}_{.\wedge t}:\gamma \sim t,\ t\geq\delta,\ N^{\gamma}_t\neq\Delta\}) = K_\delta$$

is compact in C(E) (being the standard part of an internal set of nearstandard points). Lemma 4.1 of Perkins (1990) now holds (and the same proof applies) with K_δ in place of $\mathbb{R}^d$. The existence of left-hand limits for S^H_t now follows on (δ,∞), and therefore on $(0,\infty)$, as in Proposition 4.2 of Perkins (1990). ∎

Remark 8.14. The discerning reader may have noticed that we have not extended part (c) of Theorem 1.4 of Perkins (1990). That is, we have not shown that under the hypotheses of Theorem 8.1 Q_{0,m_0}-a.s. for all $t>0$ $\text{card}(S^H_{t-}-S^H_t) = 0$ or 1 and $\text{card}(S^X_{t-}-S^X_t) = 0$ or 1. This is in fact true and it is not hard to adapt the proof given in Perkins (1990) to the current setting. Indeed it is an interesting exercise to re-interpret the nonstandard objects introduced in the proof in terms of the historical process. A very elegant proof of $\text{card}(S^X_{t-}-S^X_t) = 0$ or 1 for all $t>0$ a.s. for a reasonable class of diffusions may be found in Le Gall (1989).

If X_t is super-Brownian motion in $\mathbb{R}^d$ $(d\geq2)$, is $S(X_t)$ totally disconnected a.s.? In spite of recent progress by Roger Tribe (1989), this question remains unresolved. Theorems 8.10(b) and 8.11 establish a much easier result, namely the connectedness of the "range" of X (although this terminology was used to describe a slightly different set in [DIP]).

Theorem 8.15. Assume the hypotheses of Theorem 8.1 and let Q_{m_0}

denote the law of the $(Y,-\sigma^2\lambda^2/2)$-superprocess, X. If $S(m_0)$ is connected (respectively path-connected), then Q_{m_0}-a.s $\bigcup_{0\le t\le T} S(X_t)$ is connected (respectively path connected) for all $T\in[0,\infty]$.

Proof. Fix ω outside the Q_{0,m_0}-null set on which the conclusions of Theorem 8.10(b) and Proposition 8.11 fail. Fix $T\in[0,\infty]$ and let $x_i\in S(X_{u_i})$ where $u_i\in[0,T]$ $i=1,2$. By Theorem 8.10(b) $x_i=y_i(u_i)$ for some $y_i\in S(H_{u_i})$ $i=1,2$, and by Proposition 8.11 each y_i is a continuous path in $\bigcup_{0\le t\le T} S(X_t)$ connecting $y_i(0)\in S(m_0)$ to $x_i=y_i(u_i)$. Hence x_i and $y_i(0)$ are in the same path-connected component of $\bigcup_{0\le t\le T} S(X_t)$. The conclusion is therefore obvious from the hypothesis on $S(m_0)$. ∎

It is natural to ask if the strong continuity results of this section remain valid for any diffusion Y or if it is necessary to assume the additional conditions (8.1) and (8.2). If Y has jumps the results are known to fail badly. For example if Y is a d-dimensional symmetric stable process of index $\alpha\in(0,2)$ then (Perkins (1990, Corollary 1.6)) $S(X_t)=\emptyset$ or $\mathbb{R}^d$ for all $t>0$. The following example shows similar behavior is possible even if Y is a Feller diffusion. It does not, however, show the condition given in Remark 8.2(c) is close to optimal.

Example 8.16 Let $\delta\in(0,1)$ and $f(t)=((\log 1/t)\vee 1)^{-\delta}$ $(t\ge 0)$. Let B_t be a 1-dimensional Brownian motion starting at x under P^x and let Y_t be the $[0,\infty)\times\mathbb{R}$-valued continuous Markov process whose laws, $P^{(s,x)}$, on the canonical space of continuous $[0,\infty)\times\mathbb{R}$-valued paths are given by

$$P^{(s,x)}(Y.\in A) = P^x((s+.,B(f(s+.)-f(s))\in A).$$

Y is the space-time process associated with the time-inhomogeneous diffusion $B(f(t))$. It is easy to check that Y is a continuous Feller process. Let $B_t(x,r) = \{t\}\times\{y\in\mathbb{R}: |y-x|\le r\} = \{t\}\times B(x,r)$. Let X_t denote the $(Y,-\lambda^2/2)$-superprocess starting at $\delta_0 = \delta_{(0,0)}$ under Q_{δ_0}. If $x\in\mathbb{R}$ and $r>0$ are fixed, then Lemma 1.3 of Evans-Perkins (1990) implies

(8.23) $$Q_{\delta_0}(X_t(B_t(x,r))>0) \ge (1+t/P^0(B_{f(t)}\in B(x,r)))^{-1}.$$

If $f(t) < r^2\wedge 1$ then

$$P^0(B(f(t))\in B(x,r))/t \ge \exp\{-(|x|+r)^2(\log 1/t)^\delta/2\}/t.$$

Use this estimate in (8.23) to conclude that for $n\ge n_0(r)$

$$Q_{\delta_0}(X_{2^{-n}}(B_{2^{-n}}(x,r))=0) \le \exp\{-(n\log 2 - n^\delta(\log 2)^\delta(|x|+r)^2/2)\}$$

and hence by Borel-Cantelli, if $\Pi:[0,\infty)\times\mathbb{R} \longrightarrow \mathbb{R}$ is the projection mapping,

$$(\Pi(\bigcup_{0<t\le\varepsilon} S(X_t)))\cap B(x,r)\ne\emptyset \text{ for all } x\in\mathbb{Q},\ r>0 \text{ and } \varepsilon>0,\ Q_{\delta_0}\text{-a.s.}$$

This implies

(8.24) $$\overline{\Pi(\bigcup_{0<t\le\varepsilon} S(X_t))} = \mathbb{R} \text{ for all } \varepsilon>0,\ S(X_0) = \{(0,0)\} \quad Q_{\delta_0}\text{-a.s.}$$

A slightly fancier conditioning argument (e.g. as in the proof of Theorem 5.2 of Evans-Perkins (1990)) shows that if $T = e^{-1}$, $g(t) = 1-\left((\log \frac{1}{T-t}) \vee 1\right)^{-\delta}$ $(\delta\in(0,1))$, Y is the space-time process associated with $B(g(t))$ and X is the $(Y,-\lambda^2/2)$-superprocess, then

(8.25) $$S(X_T) = \emptyset \text{ or } \{T\}\times\mathbb{R} \quad Q_{\delta_0}\text{-a.s.}$$

(8.24) shows that the supports of X propagate instantaneously and results such as Theorems 8.10(c) and 8.12 fail. A simple calculation shows Theorem 8.1 also fails to hold, i.e. if (8.3) holds for some $G\in\mathcal{H}$, then $\delta_M \xrightarrow{w} 0$ and so (8.4) must fail.

Appendix.

1. Proofs for Section 2.

We give proofs of many of the routine results in Section 2. The notation is that introduced in Section 2.

Lemma A.1. Let $s\geq 0$. If $\hat{T}$ is a $(\hat{\mathcal{D}}_{t+})$-stopping time, then $T(y) = s+\hat{T}(\hat{W}_s(y))$ is a $(\mathcal{D}_{t+})$-stopping time. If $\hat{A}\in \hat{\mathcal{D}}_{T+}$, then

$$A = \{y\in D(E):\hat{W}_s(y)\in \hat{A}\} \in \mathcal{D}_{T+}.$$

Proof. Let $\hat{T},T,\hat{A}$ and A be chosen as above. We claim

(A.1) $A\cap\{T<u\} \in \mathcal{D}_u \quad \forall\ u\geq s$ (and hence all $u\geq 0$).

If $u\geq s$,

$y\in A$ and $T(y)<u \iff \hat{W}_s(y)\in \hat{A}$ and $\hat{T}(\hat{W}_s(y)) < u-s$

(A.2) $\iff \hat{W}_s(y)^{u-s}\in \hat{A}$ and $\hat{T}(\hat{W}_s(y)^{u-s}) < u-s$

(by [DM, IV.96(c)] since $\hat{A}\cap\{\hat{T}<u-s\}\in\hat{\mathcal{D}}_{u-s}$)

$\iff \hat{W}_s(y^u)^{u-s}\in \hat{A}$ and $\hat{T}(\hat{W}_s(y^u)^{u-s}) < u-s$

(definition of $\hat{W}_s$)

$\iff \hat{W}_s(y^u)\in \hat{A}$ and $\hat{T}(\hat{W}_s(y^u)) < u-s$ (as in (A.2) with y^u in place of y)

$\iff y^u \in A$ and $T(y^u) < u$.

(A.1) now follows from the above equivalence and a further application of [DM, IV.96(c)]. Take $\hat{A} = D(\hat{E})$ to see T is a $(\mathcal{D}_{t+})$-stopping time. $A\in \mathcal{D}_{T+}$ is now immediate from (A.1). ∎

Proof of Proposition 2.1.2. (H_1) implies $(s,y) \longrightarrow \hat{P}^{(s,y)}(A)$ is $\hat{\mathcal{E}}$-measurable $\forall\ A\in \hat{\mathcal{D}}$. (H_2) implies $\hat{P}^{(s,y)}(\hat{Y}(0)=(s,y)) = 1$ $\forall\ (s,y)\in \hat{E}$. Let $\hat{T}$ be a $(\hat{\mathcal{D}}_{t+})$-stopping time and let $\hat{A}\in \hat{\mathcal{D}}_{T+}$, $\hat{A} \subset \{\hat{T}<\infty\}$. Define A and T as in Lemma A.1. If $f\in b\hat{\mathcal{E}}$, then

$$\int 1_{\hat{A}} \hat{P}_t f(\hat{Y}(\hat{T}))d\hat{P}^{(s,y)} = \int 1_{\hat{A}}(\hat{W}_s)\hat{P}_t f(\hat{W}_s(\hat{T}(\hat{W}_s)))\ dP_{s,y}$$

$$= \int 1_A \hat{P}_t f(T,W(T))\ dP_{s,y}$$

$$= \int 1_A P_{T(\omega),W(T)(\omega)}(f(t+T(\omega),W(t+T(\omega)))\ dP_{s,y}$$

(by (2.1.1)

$$= \int 1_A f(t+T,W(t+T))\ dP_{s,y} \qquad \text{(from } (H_3))$$

$$= \int 1_{\hat{A}}(\hat{W}_s) f(s+t+\hat{T}(\hat{W}_s),W(s+t+\hat{T}(\hat{W}_s)))\ dP_{s,y}$$

$$= \int 1_{\hat{A}}(\hat{W}_s) f(\hat{W}_s(t+\hat{T}(\hat{W}_s)))dP_{s,y}$$

$$= \int 1_{\hat{A}} f(\hat{Y}(t+\hat{T}))\ d\hat{P}^{(s,y)}.$$

We have proved

$$\hat{P}^{(s,y)}(f(\hat{Y}(t+\hat{T}))|\hat{\mathcal{D}}_{\hat{T}+}) = \hat{P}_t f(\hat{Y}(\hat{T})) \quad \hat{P}^{(s,y)}\text{-a.s. on } \{\hat{T}<\infty\}.$$

Since $\hat{P}_t : b\hat{\mathcal{E}} \longrightarrow b\hat{\mathcal{E}}$ we may apply Theorem (9.4)(i) of Getoor (1979) to conclude that $\hat{Y}$ is a Borel right process.

Assume W is an inhomogeneous Hunt process and let $\{\hat{T}_n\}$ be $(\hat{\mathcal{D}}_{t+})$-stopping times which increase to $\hat{T}$ and assume $\hat{T} < \infty$ $\hat{P}^{(s,y)}$ a.s. Let $T_n = s+\hat{T}_n(\hat{W}_s)$ and $T = s+\hat{T}(\hat{W}_s)$. By Lemma A.1 and the quasi-left-continuity of W we have

$$W(T_n) \longrightarrow W(T) \quad P_{s,y}\text{-a.s.}$$

$$\Longrightarrow (s+\hat{T}_n(\hat{W}_s),W(s+\hat{T}_n(\hat{W}_s)) \longrightarrow (s+\hat{T}(\hat{W}_s),W(s+\hat{T}(\hat{W}_s))) \quad P_{s,y}\text{-a.s.}$$

$$\Longrightarrow \hat{W}_s(\hat{T}_n(\hat{W}_s)) \longrightarrow \hat{W}_s(\hat{T}(\hat{W}_s)) \quad P_{s,y}\text{-a.s.}$$

$$\Longrightarrow \hat{Y}(\hat{T}_n) \longrightarrow \hat{Y}(\hat{T}) \quad \hat{P}^{(s,y)}\text{-a.s.} \blacksquare$$

We now turn to the proof of Theorem 2.1.5. The proof of the next lemma is similar to that of A.1 and is therefore omitted. The σ-fields $\hat{\mathcal{G}}_{t+}$ and $\mathcal{G}[s,t]$ are defined prior to Lemma 2.1.4.

Lemma A.2. Let $s\geq 0$, $T\geq s$ be a $\{\mathcal{G}[s,t+]:t\geq s\}$-stopping time and assume $A\in \mathcal{G}[s,T+]$. Then $\hat{T}(\omega) = T(\hat{H}^{(s)}(\omega))-s$ is a

$\{\hat{\mathcal{G}}_{t+}\}$-stopping time and $\hat{A} = \{\omega:\hat{H}^{(s)}(\omega)\in A\} \in \hat{\mathcal{G}}_{\hat{T}+}$.

Proof of Theorem 2.1.5. (a) The Borel measurability of $\nu \longrightarrow \hat{Q}_\nu(B)$ $(B\in\mathcal{G})$ on $M_F(\hat{E})$, implies the Borel measurability of $(\nu,t) \longrightarrow \hat{Q}_\nu(\hat{H}\circ\hat{\theta}_t\in A)$ $(A\in \mathcal{G})$. If u is fixed then $(A\in \mathcal{G})$

$$Q_{s,m}(H\circ\theta_u\in A) = \hat{Q}_{\delta_s\times m}(\hat{H}\circ\hat{\theta}_{u-s}\in A)$$

is therefore Borel measurable on $([0,u]\times M_F(D))\cap \hat{M}$. Lemma 2.1.4 implies that $H_t\in M_F(D)^t$ $\forall t\geq s$ $Q_{s,m}$-a.s. Since $Q_{s,m}(H_s=m) = 1$ is obvious it remains only to check the strong Markov property.

Fix $(s,m)\in \hat{M}$, let $T\geq s$ be a $(\mathcal{G}[s,t+])$-stopping time, $A\in \mathcal{G}[s,T+]$ with $A\subset\{T<\infty\}$, and let $\psi\in b\mathcal{G}$. Define $\hat{T}$ and $\hat{A}$ as in Lemma A.2. Then

$$\begin{aligned}
\int 1_A\psi(H\circ\theta_T)\,dQ_{s,m} &= \int 1_A(\hat{H}^{(s)})\psi(\hat{H}^{(s)}(T(\hat{H}^{(s)})+.))\, d\hat{Q}_{\delta_s\times m} \\
&= \int 1_{\hat{A}}\ \psi(\hat{H}(\hat{T}+.))\,d\hat{Q}_{\delta_s\times m} \\
&= \int 1_{\hat{A}}\hat{Q}_{\hat{X}(\hat{T})}(\psi(\hat{H}))\,d\hat{Q}_{\delta_s\times m} \qquad \text{(by the SMP for } \hat{X}) \\
&= \int 1_{\hat{A}}\hat{Q}_{\delta_{s+\hat{T}}\times\hat{H}(\hat{T})}(\psi(\hat{H}))\,d\hat{Q}_{\delta_s\times m} \qquad \text{(Lemma 2.1.4(a))} \\
&= \int 1_A(\hat{H}^{(s)})\hat{Q}_{\delta_{T(\hat{H}^{(s)})}\times\hat{H}^{(s)}(T(\hat{H}^{(s)}))}(\psi(\hat{H}))\,d\hat{Q}_{\delta_s\times m} \\
&= \int 1_A\hat{Q}_{\delta_T\times H(T)}(\psi(\hat{H}))\,dQ_{s,m} \\
&= \int 1_A(\omega)Q_{T(\omega),H(T)(\omega)}(\psi(H\circ\theta_{T(\omega)}))\ dQ_{s,m}.
\end{aligned}$$

This gives the strong Markov property ((iv) of Defintion 2.1.0) and (a) is proved.

(b) Let $T_n\geq s$ be $\{\mathcal{G}[s,t+]:t\geq s\}$-stopping times which increase to T satisfying $Q_{s,m}(T<\infty) = 1$. Define $\hat{T}_n$ and $\hat{T}$ as in Lemma A.2. Note that $\hat{Q}_{\delta_s\times m}(\hat{T}<\infty) = 1$. Prop. 2.1.2 and Theorem 2.1.3 imply $\hat{X}$ is quasi-left-continuous and hence $\hat{X}(\hat{T}_n) \longrightarrow \hat{X}(\hat{T})$, $\hat{Q}_{\delta_s\times m}$-a.s. This implies $\hat{Q}_{\delta_s\times m}$-a.s.

$$\hat{H}(T_n(\hat{H}^{(s)})-s) \longrightarrow \hat{H}(T(\hat{H}^{(s)})-s)$$
$$\Longrightarrow \hat{H}^{(s)}(T_n(\hat{H}^{(s)})) \longrightarrow \hat{H}^{(s)}(T(\hat{H}^{(s)}))$$

and therefore $H(T_n) \longrightarrow H(T)$, $Q_{s,m}$-a.s., as required.

(c) Theorem 2.1.3 and Proposition 2.1.2 show $\hat{X}.$ is a.s. continuous if W is a Hunt process and $\hat{n} = 0$. The a.s.-continuity of H follows.

(d)
$$\begin{aligned} Q_{s,m}(\langle H_t, f\rangle) &= \hat{Q}_{\delta_s\times m}(\langle \hat{H}_{t-s}, f\rangle) \\ &= \langle m, \hat{P}^{\hat{b}}_{t-s}\hat{f}(s,.)\rangle \quad \text{by (2.1.7)}, \\ &\qquad\qquad \text{where} \quad \hat{f}(u,y) = f(y) \\ &= \langle m, T^{\hat{b}}_{s,t} f\rangle, \end{aligned}$$

the last by an elementary calculation from the definition of $\hat{p}^{(s,y)}$. ∎

2. <u>Measurability Lemmas for Section 5.</u>

<u>Lemma A.3.</u> Let $\psi\in\mathcal{G}$ and let Λ be a G_δ in $\mathbb{R}^d\times M_F(\mathbb{R}^d)$. If $B \subset \mathbb{R}^d$ is compact and $\Lambda(\nu) = \{x:(x,\nu)\in \Lambda\}$, then $\nu \longrightarrow \psi\text{-m}(\Lambda(\nu)\cap B)$ is universally measurable on $M_F(\mathbb{R}^d)$. In fact $\{\nu:\psi\text{-m}(\Lambda(\nu)\cap B) > \lambda\}$ is an analytic subset of $M_F(\mathbb{R}^d)$ $\forall\ \lambda\geq 0$.

<u>Proof.</u> The space $\mathcal{K}$ of compact subsets of B with the Hausdorff metric topology is a compact metric space (see Dugundji (1966, p. 205, 253)). The inner regularity of ψ-m (Rogers (1970, Thms. 47,48)) implies

(A.3) $\{\nu:\psi\text{-m}(\Lambda(\nu)\cap B) > \alpha\} = \Pi(\Lambda')$

where $\Lambda' = \{(K,\nu)\in \mathcal{K}\times M_F(\mathbb{R}^d):K\subset \Lambda(\nu),\ \psi\text{-m}(K) > \alpha\}$ and $\Pi:\mathcal{K}\times M_F(\mathbb{R}^d) \longrightarrow M_F(\mathbb{R}^d)$ is the projection mapping. Assume $\Lambda = \bigcap_n \Lambda_n$ where Λ_n are open subsets of $\mathbb{R}^d\times M_F(\mathbb{R}^d)$. Then

(A.4) $\Lambda' = (\bigcap_n \Lambda_n')\cap\Lambda''$,

where $\Lambda_n' = \{(K,\nu)\in \mathcal{K}\times M_F(\mathbb{R}^d):K \subset \Lambda_n(\nu)\}$

and

$$\Lambda'' = \{(K,\nu)\in \mathcal{K}\times M_F(\mathbb{R}^d):\psi\text{-m}(K) > \alpha\}.$$

We claim Λ'_n is open. Let $(K,\nu)\in \Lambda'_n$, so that $K\times\{\nu\} \subset \Lambda_n$. Since Λ_n is open there is a $\delta > 0$ such that $\text{int}(K^\delta)\times \text{int}(B(\nu,\delta)) \subset \Lambda_n$, where $B(\nu,\delta)$ is the closed ball in $M_F(\mathbb{R}^d)$ with respect to an appropriate metric d. Therefore

$$U = \{(K',\nu')\in \mathcal{K}\times M_F(\mathbb{R}^d):K' \subset \text{int}(K^\delta),\ d(\nu',\nu) < \delta\} \subset \Lambda'_n.$$

U is an open neighbourhood of (K,ν) (Theorem 4.2.3 of Cutler (1984)) and so Λ'_n is open. An easy modification of Theorem 4.5.1 of Cutler (1984) shows that $K \longrightarrow \psi\text{-m}(K)$ is Borel measurable on $\mathcal{K}$. Therefore Λ' is Borel measurable (by (A.4)) and so $\{\nu:\psi\text{-m}(\Lambda(\nu)\cap B) > \alpha\}$ is an analytic subset of $M_F(\mathbb{R}^d)$ by (A.3) ([DM,III.9]). ∎

Recall $\Gamma(\nu,\psi,c) = \{x: \overline{\lim_{r\downarrow 0}}\ \nu(B(x,r))/\psi(r) \geq c\}$.

<u>Corollary A.4.</u> Let $\psi\in \mathcal{I}$ and B be a compact subset of $\mathbb{R}^d$. Then the mapping $\nu \longrightarrow \psi\text{-m}(\Gamma(\nu,\psi,c)\cap B)$ defined on $M_F(\mathbb{R}^d)$ is universally measurable $\forall$ $c\geq 0$.

<u>Proof.</u> Fix $c > 0$ and let

$$\Gamma = \{(x,\nu)\in \mathbb{R}^d\times M_F(\mathbb{R}^d): \overline{\lim_{r\downarrow 0}}\ \nu(B(x,r))/\psi(r) \geq c\}.$$

If

$$\Gamma_{m,n} = \bigcup_{0<r\leq 1/m} \{(x,\nu):\nu(\text{int}(B(x,r)))/\psi(r) > c - 1/n\}$$

then $\Gamma = \bigcap_m \bigcap_n \Gamma_{m,n}$. It is easy to see that

(A.5) $$\underline{\lim}_{k\to\infty}\ \nu_k(\text{int}\ B(x_k,r)) \geq \nu(\text{int}(B(x,r))) \quad \text{whenever} \quad (x_k,\nu_k) \longrightarrow (x,\nu).$$

Therefore, being the union of open sets, $\Gamma_{m,n}$ is open, Γ is a G_δ and the result follows from Lemma A.3.

<u>Remark.</u> It is also easy to use Lemma A.3 to show $\nu \longrightarrow \psi\text{-m}(S(\nu)\cap B)$ is universally measurable on $M_F(\mathbb{R}^d)$ (ψ and B as in A.4). For this note that if $S = \{(x,\nu):x\in S(\nu)\}$ and $S_n =$

$\{(x,\nu):\nu(\text{int}(B(x,1/n)) > 0\}$ then S_n is open by (A.5) and $S = \bigcap_n S_n$. In fact Lemma 6.3 of [DIP] shows that this map is Borel measurable.

3. Complement to Section 3.

Proof of Lemma 3.2. We have for $\phi \in bp\mathcal{E}_1 \times \mathcal{E}_2$ and conditioned on $\mathcal{G}_1$

$$E\left[e^{-\langle X,\phi\rangle} | \mathcal{G}_1\right] = \exp\left(\iint -(1 - e^{-\int \phi(x,y)\mu(dy)})P_x(d\mu)I(dx)\right).$$

Let $X^*(A) = X(\tilde{A})$, $A \in \mathcal{E}_1$, $\tilde{A} = A \times E_2$ and $\psi \in bp\mathcal{E}_1$. Then

$$E\left[e^{-\langle X^*,\psi\rangle} | \mathcal{G}_1\right] = E\left[e^{-\langle X,\phi\rangle} | \mathcal{G}_1\right] \quad \text{with} \quad \phi(x,y) = \psi(x)$$

$$= \exp\left(-\int (1-e^{-\psi(x)\langle\mu,1\rangle})P_x(d\mu)I(dx)\right).$$

Hence from (3.4) X^* is a Poisson cluster process with intensity I and cluster law $P_x^*(B) = \int 1_B(a\delta_x)P_x(\langle\mu,1\rangle \in da)$, $B \in \mathcal{B}(M_F(E_2))$.

Since I is a.s. a finite measure, X^* is a.s. of the form $\sum_{i=i}^{N} Y_i\delta_{y_i}$ where N is finite, $\{Y_i\}$ are independent positive random variables random variables, Y_i has distribution $P_{y_i}(\langle\mu,1\rangle \in .)$ and $\tilde{X} = \sum_{i=1}^{N} \delta_{y_i}$ is a Poisson random field on E_1 with intensity I.

It then follows by an elementary argument that a.s. on $\{I\text{:}I \text{ non-atomic}\}$, $\{y_i\}$ are distinct points and

$$\tilde{X}(G) = \lim_{n\to\infty} \sum_{i=1}^{\infty} 1(A_i^n \subset G)1_{(0,\infty)}(X^*(A_i^n)) \quad \text{for } G \text{ open}.$$

Hence $\tilde{X}$ is $\mathcal{G}_0$-measurable.

If $A \in \mathcal{E}_1$, then

$$E\left(\mathbf{1}(X(\tilde{A})=0)e^{-<X,\phi>}|\mathcal{G}_1\right) = \lim_{\theta\to\infty} E\left(e^{-<X,\phi>+\theta<X,\mathbf{1}_{\tilde{A}}>}|\mathcal{G}_1\right)$$

$$= \lim_{\theta\to\infty} \exp\left(-\iint(1 - e^{-\int[\phi(x,y)+\theta\mathbf{1}_A(x)]\mu(dy)})P_x(d\mu)I(dx)\right)$$

$$= \exp\left(-\int_{A^c}\int(1 - e^{-\int\phi(x,y)\mu(dy)})P_x(d\mu)I(dx)\right).e^{-I(A)}.$$

On the other hand,

$$E\left(\mathbf{1}(X(\tilde{A})=0)\exp\left(\int\log\left(\int e^{-\int\phi(x,y)\mu(dy)}P_x(d\mu)\right)\tilde{X}(dx)\right)|\mathcal{G}_1\right)$$

$$= E\left(\mathbf{1}(X(\tilde{A})=0)\exp\left(\int_{A^c}\log\left(\int e^{-\int\phi(x,y)\mu(dy)}P_x(d\mu)\right)\tilde{X}(dx)\right)|\mathcal{G}_1\right)$$

(since $E[\tilde{X}(A)\mathbf{1}(X(\tilde{A})=0)] = 0$)

$$= \lim_{\theta\to\infty} E\left(\exp\left(-\theta\tilde{X}(A) - \int_{A^c}-\log\left(\int e^{-\int\phi(x,y)\mu(dy)}P_x(d\mu)\right)\tilde{X}(dx)\right)|\mathcal{G}_1\right)$$

(since $X(\tilde{A}) = 0 \Leftrightarrow \tilde{X}(A) = 0$)

$$= \lim_{\theta\to\infty} \exp\left(-[1- e^{-\theta}]I(A) -\int_{A^c}\left[1 - \int e^{-\int\phi(x,y)\mu(dy)}P_x(d\mu)\right]I(dx)\right)$$

(by (3.3))

$$= \exp\left(-\int_{A^c}\int(1 - e^{-\int\phi(x,y)\mu(dy)})P_x(d\mu)I(dx)\right).e^{-I(A)}.$$

Hence for $A\in\mathcal{E}_1$

$$E\left(\mathbf{1}(X(\tilde{A})=0)e^{-<X,\phi>}|\mathcal{G}_1\right)$$

$$= E\left(\mathbf{1}(X(\tilde{A})=0)\exp\left(\int\log\left(\int e^{-\int\phi(x,y)\mu(dy)}P_x(d\mu)\right)\tilde{X}(dx)\right)|\mathcal{G}_1\right).$$

Since $\tilde{X}$ is $\mathcal{G}_0$-measurable and $\mathcal{G}_0 = \sigma\{X(\bar{A})=0: A\in \mathcal{E}_1\}$, a.s.

$$E\left(e^{-\langle X,\phi\rangle}|\mathcal{G}_0 \vee \mathcal{G}_1\right) = \exp\left(\int \log\left(\int e^{-\int \phi(x,y)\mu(dy)} P_x(d\mu)\right)\tilde{X}(dx)\right).$$

Finally, replacing ϕ by $\theta\phi$, differentiating with respect to θ and evaluating at $\theta=0$ yields

$$E(\langle X,\phi\rangle|\mathcal{G}_0 \vee \mathcal{G}_1) = \int\int\int \phi(x,y)\mu(dy)P_x(d\mu)\tilde{X}(dx) \quad \text{a.s.}$$

and the proof is complete. ∎

Let S be a Polish space and let d be a metric for the Skorohod topology on $D(S)$.

Lemma A.5. Let $(X;X_n, n\in\mathbb{N})$ denote $D(S)$-valued random variables defined on a common probability space. Assume that

(i) the laws of $\{X_n\}$ are relatively compact in $D(S)$

and

(ii) $X_n(t) \underset{n\to\infty}{\longrightarrow} X(t)$ in probability for each t in a dense subset of $[0,\infty)$.

Then $d(X_n,X) \longrightarrow 0$ in probability as $n \to \infty$.

Proof. Assume the contrary. Then $\exists\ \varepsilon > 0$ such that $\forall\ k\ \exists n_k, m_k > k$ such that

$$P(d(X_{n_k}, X_{m_k}) \geq \varepsilon) \geq \varepsilon.$$

But $\{(X_{n_k}, X_{m_k}): k\in\mathbb{N}\}$ is tight in $D(S)\times D(S)$ by (i). Hence there exists a weakly convergent subsequence $(X_{n_{k'}}, X_{m_{k'}})$. By (ii) the limiting law must coincide with that of $\{(X,X)\}$ (cf. Ethier Kurtz (1986, Chap. 3, Thm. 7.8)). Hence

$$\lim_{k'\to\infty} P(\ d(X_{n_{k'}}, X_{m_{k'}}) < \varepsilon\ ) = 1$$

yielding a contradiction and the proof is complete. ∎

4. Complement to Section 6.

Let μ be a finite measure on a Polish space E and

$\ell a(\mu) = x$ if $\mu(\{x\})\delta_x$ is the atom of largest mass of μ (with no ties)

$= \emptyset$ otherwise.

Given $\mu \in M_p^t$ let $\mu_p(A) = \int_A \phi_{p,\infty}(w)\mu(dw) \in M_F(\bar{D}^{\pm})$.

Lemma A.6.

(a) $\mu \in M_p^t$ is a clan measure if and only if

$$\text{(A.6)} \quad \lim_{s\downarrow -\infty} \mu_p(\{w:w^s=\ell a(r_s\mu_p)\}) = \mu_p(\bar{D}^{\pm,t}).$$

(b) $M_p^{cl,t}$ is a measurable subset of M_p^t.

Proof. (a) If μ is a clan measure supported by a clan C (see definition immediately before Theorem 6.3), then for y_0 as in the above definition, $\ell a(r_s\mu_p) = y_0^s$ for sufficiently large negative s and so

$$\lim_{s\downarrow -\infty} \mu_p(\{w:w^s=\ell a(r_s\mu_p)\})$$

$$= \lim_{s\downarrow -\infty} \int \phi_{p,\infty}(w)\mathbf{1}(w\in C,\tau(w)\geq s)\mu(dw) = \mu_p(\bar{D}^{\pm}).$$

On the other hand if μ satisfies (A.6), then there exists s for which $\ell a(r_s\mu_p) = y \in \bar{D}^{\pm,s}$ exists. But it is then easy to check that μ is supported by the clan

$$C = \{w:w^u=y^u \text{ for some } u\leq s\}.$$

(b) The mapping $\mu \longrightarrow \mu_p(\bar{D}^{\pm})$ is continuous and therefore measurable on M_p^t. Hence it suffices to show that the mapping

$$\mu \longrightarrow \mu_p(\{w:w^s=\ell a(r_s\mu_p)\})$$

is measurable from M_p^t to $\mathbb{R}$. But this follows since the mapping $\mu \longrightarrow \ell a(r_s\mu_p)$ is measurable from M_p^t to $\bar{D}^{\pm,s}$ and the mapping

$(y,\mu) \longrightarrow \mu_p(\{w:w^s=y\})$ is measurable from $\bar{D}^{\pm,s}\times M_p^t$ to $\mathbb{R}$ (cf. Kallenberg (1977b, Lemma 2.3)). ∎

Let $f_R(x) = 1_{B(0,R)}(x)$ and $u_1(t,x;\theta) = U_t(\theta f_R)$ denote the solution of (6.1a) with $\phi = \theta f_R$.

Lemma A.7. If $\alpha = 2$ then

$$\lim_{t\to\infty} \int \lim_{\theta\to\infty} u_1(t,x;\theta)dx < \infty.$$

Proof. The following is a slight modification of the proof of Lemma 3.1 in [DIP] (there β=1). The mapping $\theta \longrightarrow u_1(t,x;\theta)$ is increasing and by (3.24)

$$\text{(A.7)} \qquad u_1(t,x;\theta) \le \frac{\theta}{(1+t\beta\theta^\beta/2)^{1/\beta}} \le (2/t\beta)^{1/\beta}.$$

Let $\lambda > 0$, and $u_2(t,x;\theta)$ denote the (non-negative) solution of

$$\frac{\partial u_2}{\partial t} = \frac{1}{2}\left(\Delta u_2 - \lambda u_2 - u_2^{1+\beta}\right), \quad u_2(0) = \theta f_R.$$

If $v= e^{\lambda t/2}.u_2$, then

$$\frac{\partial v}{\partial t} = (\Delta v - e^{-\lambda t\beta/2}\, v^{1+\beta})/2 \ge (\Delta v - v^{1+\beta})/2, \quad v(0) = \theta f_R.$$

Let $g(v,u) = \begin{cases} (v^{1+\beta}-u^{1+\beta})/(2(v-u)) & \text{if } u\ne v \\ (1+\beta)v^\beta/2 & \text{if } u=v \end{cases}$, and define

$c(t,x) = g(v(t,x),u_1(t,x))$. Then $c\ge 0$ is continuous and

$$\frac{\partial(v-u_1)}{\partial t} \ge \Delta(v-u_1)/2 - (v^{1+\beta}-u_1^{1+\beta}) = \Delta(v-u_1)/2 - c(v-u_1),$$

$$(v-u_1)(0)=0.$$

Hence by a comparison lemma [DIP , Lemma 3.0] with $\mathcal{D} \equiv \mathbb{R}^d$, we obtain $u_1 \le v$. In turn by [DIP, Lemma 3.0] with $\mathcal{D} \equiv \overline{B(0,R+\delta)}^c$

(δ being sufficiently small and positive but otherwise arbitrary), $u \equiv u_3 - u_2$ and $c = \lambda/2 + g(u_3,u_2)$ we obtain that $u_2 \le u_3$ on $\overline{B(0,R)}^c$, where u_3 is any positive solution of

$$-\Delta u_3 + \lambda u_3 + u_3^{1+\beta} \ge 0, \quad |x| > R$$

$$u_3(x) \longrightarrow \infty \text{ as } |x| \longrightarrow R+$$

$$u_3(x) \longrightarrow 0, \text{ as } |x| \longrightarrow \infty.$$

In radial form (with $'$ denoting differentiation with respect to r) this becomes

$$-u_3''(r) - \frac{(d-1)}{r} u_3'(r) + \lambda u_3(r) + u_3^{1+\beta} \ge 0, \; r > R$$

$$u_3(r) \longrightarrow \infty \quad \text{as} \quad r \longrightarrow R+$$

$$u_3(r) \longrightarrow 0 \text{ as } r \longrightarrow \infty.$$

For example one can choose $u_3(r) = w(r-R)$ such that w is positive and

$$w'' = \lambda w + w^{1+\beta}, \quad r > 0$$

$$w' < 0, \; r > 0$$

$$w(r) \longrightarrow \infty \text{ as } r \longrightarrow 0+$$

$$w(r) \longrightarrow 0 \text{ as } r \longrightarrow \infty.$$

The solution is given by

$$w(r) = (\lambda(2+\beta)/2)^{1/\beta}(\sinh(\sqrt{\lambda}\beta r/2))^{-2/\beta}$$

as is easily verified. Letting $\lambda = 1/t$ we obtain from (A.7) and the above

(A.8) $$u_1(t,x;\theta)$$

$$\le \left\{t^{-1}.\min\left(2\beta^{-1}, e^{\beta/2}(1+\beta/2)\sinh(\beta(|x|-R)^+/2\sqrt{t})^{-2}\right)\right\}^{1/\beta}.$$

Now let $u_\infty(t,x) = \lim_{\theta\to\infty} u_1(t,x;\theta)$. By (A.8) $u_\infty(t,.)$ is Lebesgue integrable and the result follows since $\int u_\infty(t,x)\,dx$ is monotone decreasing in t (see the argument prior to Proposition 6.1). ∎

<u>Corollary A.8.</u> Assume the hypotheses of Proposition 6.10. Then

$$R_\infty(\{\mu:\mu(B(0,1))>0\}) < \infty.$$

<u>Proof.</u> If $u_1(t,x;\theta)$ is as in Lemma A.7 with $R=1$ and $f = 1_{B(0,1)}$, then (6.25) implies

$$\begin{aligned} R_\infty(\{\mu:\mu(B(0,1))>0\}) &= \lim_{\theta\to\infty} \int (1-e^{-\langle\mu,\theta f\rangle}) R_\infty(d\mu) \\ &= \lim_{\theta\to\infty} \lim_{t\to\infty} \int u_1(t,x;\theta)\,dx \\ &\le \lim_{t\to\infty} \int u_\infty(t,x)\,dx \end{aligned}$$

since $u_1(t,x;\theta)$ increases to $u_\infty(t,x)$. The above limit is finite by Lemma A.7. ∎

<u>Lemma A.9.</u> (a) Let $\hat{Q}_{s,\xi_s}$ be defined by (6.14a) with $\xi_s \in M_p^{cl,s}$. Then

$$\hat{Q}_{s,\xi_s}(\xi_t \in M_p^{cl,t}\ \forall t\ge s) = 1.$$

(b) Let $\tilde{H}_t$ and Ξ_t^* be as in the statement and proof of Theorem 6.4(c). Then

$$\tilde{H}_t = \mathcal{J}(\Xi_t^*)\ \forall t\ge 0,\ \tilde{Q}^p\text{-a.s.}$$

<u>Proof.</u> (a) By Proposition 6.2

$$|\langle\xi_{s+\varepsilon},\phi\rangle - \langle\xi_{s+\varepsilon}^{(n)},\phi\rangle| \longrightarrow 0\ \forall\phi\in C_p^+,\ \hat{Q}_{s,\xi_s}\text{-a.s.}$$

where $\xi_t^{(n)}(\cdot) = \xi_t(\cdot\cap\{y:|y(s)|\le n\})$, $t\ge s$. Let $\varepsilon>0$. By (the analogue of) Proposition 3.3(b) $\xi_{s+\varepsilon}^{(n)}$ is a Cox cluster random measure on $\bar{D}^{\pm,s+\varepsilon}$ with intensity $\xi_s(dy)1(|y(s)|\le n)/(\varepsilon\beta\gamma)^{1/\beta}$ and cluster law $P^*_{s,s+\varepsilon;y}$. Letting $n\uparrow\infty$ it follows that $\xi_{s+\varepsilon}$ has a Cox cluster decomposition

$$\xi_{s+\varepsilon} = \sum_{j=1}^{\infty} \xi_{s+\varepsilon,j} \;, \qquad \hat{Q}_{s,\xi_s}\text{-a.s.}$$

where $r_s\xi_{s+\varepsilon,j} = a_j\delta_{y_j}$ for some $a_j \in (0,\infty)$ and $y_j \in \bar{D}^{\pm,s}$ and

$\xi_{t,j}(\cdot) = \xi_t(\cdot\cap\{y:y^s=y_j\})$, $t \ge s+\varepsilon$. Since $\xi_s \in M_p^{cl,s}$, the set $\{y_j:j\in\mathbb{N}\}$ is a clan.

By (the analogue of) Proposition 3.3 it follows that

$$\xi_t = \sum_{j=1}^{\infty} \xi_{t,j} \;, \; \hat{Q}_{s,\xi_s}\text{-a.s. for } t \ge s+\varepsilon .$$

It now suffices to show that

$$\xi_t(\{y:y^s \notin \{y_j:j\in\mathbb{N}\}\}) = 0 \quad \forall t \ge s+\varepsilon \;, \; \hat{Q}_{s,\xi_s}\text{-a.s. } \forall\varepsilon>0,$$

or equivalently

$$\xi_t = \sum_{j=1}^{\infty} \xi_{t,j} \quad \forall t \ge s+\varepsilon, \; \hat{Q}_{s,\xi_s}\text{-a.s.}$$

By an argument similar to that of the proof of Proposition 6.2 we obtain for any $T \ge s+\varepsilon$,

$$\sup_{s+\varepsilon \le t \le T} |\langle\xi_t,\phi\rangle - \sum_{j=1}^{N} \langle\xi_{t,j},\phi\rangle| \longrightarrow 0 \;\; \forall\phi\in C_p^+, \;\; \hat{Q}_{s,\xi_s}\text{-a.s.}$$

and therefore

$$\sup_{s+\varepsilon \le t \le T} |\langle\xi_t,\phi\rangle - \sum_{j=1}^{\infty} \langle\xi_{t,j},\phi\rangle| = 0 \quad \forall\phi\in C_p^+, \;\; \hat{Q}_{s,\xi_s}\text{-a.s.}$$

Since ε and T are arbitrary, this completes the proof of (a).

(b) Again by an argument similar to that of the proof of Proposition 6.2 we obtain for any $T>0$,

$$\sup_{0 \le t \le T} |\langle\tilde{H}_t,\phi\rangle - \sum_{j=1}^{N} \langle\xi_t^{(j)},\phi\rangle| \longrightarrow 0 \;\; \forall\phi\in C_p^+, \;\; \tilde{Q}^p\text{-a.s.}$$

Therefore

$$\sup_{0\le t\le T} |\langle \tilde{H}_t,\phi\rangle - \langle \mathcal{J}(\Xi_t^*),\phi\rangle|$$

$$= \sup_{0\le t\le T} |\langle \tilde{H}_t,\phi\rangle - \sum_{j=1}^{\infty} \langle \xi_t^{(j)},\phi\rangle| = 0 \quad \forall\phi\in C_p^+,\ \tilde{Q}^p\text{-a.s.}$$

Since T is arbitrary, this completes the proof. ∎

References

Anderson, R.M. and Rashid, S. (1978). A nonstandard characterization of weak convergence, Proc. Amer. Math. Soc. **69**, 327-332.

Blumenthal, R.M. and Getoor, R.K. (1968). *Markov processes and potential theory*, Academic Press, New York.

Chauvin, B. (1986). Arbres et processus de Bellman-Harris, Ann. Inst. Henri Poincaré **22**, 209-232.

Cutland, N (1983). Nonstandard measure theory and its applications, Bull. London Math. Soc. **15**, 529-589.

Cutler, C. (1984). Some measure-theoretic and topological results for measure-valued and set-valued stochastic processes, Ph.D. thesis, Carleton University.

Dawson, D.A. (1977). The critical measure diffusion, Z. Wahr. verw. Geb. **40**, 125-145.

Dawson, D.A. and Ivanoff, G. (1978). Branching diffusions and random measures. In *Branching Processes*, A. Joffe and P. Ney, eds., M. Dekker, 61-104.

Dawson, D.A. (1978). Limit theorems for interaction free geostochastic systems, Colloquia Math. Soc. J. Bolyai **24**, 27-47.

Dawson, D.A. and Fleischmann, K. (1985). Critical dimension for a model of branching in a random medium, Z. Wahr. verw. Geb. **70**, 315-334.

Dawson, D.A., Fleischmann, K. and Gorostiza, L.G. (1989). Stable hydrodynamic limit fluctuations of a critical branching particle system, Ann. Probab. **17**, 1083-1117.

Dawson, D.A., Iscoe, I. and Perkins, E.A. (1989). Super-Brownian motion: path properties and hitting probabilities, Probab. Th. Rel. Fields **83** ,135-206.

Dawson, D.A. and Gorostiza, L.G. (1990). Generalized solutions of a class of nuclear space valued stochastic evolution equations, Appl. Math and Opt., to appear.

Dellacherie, C. and Meyer, P.A. (1978). *Probability and Potential*, North Holland Mathematical Studies 29, North Holland, Amsterdam.

Donnelly, P. and S. Tavaré (1987) The population genealogy of the infinitely-many neutral alleles model, J. Math. Biol. **25**, 381-391.

Donnelly, P. and T.G. Kurtz (1989). Infinite population genetic models, preprint.

Dugundji, J. (1966). *Topology*. Allyn and Bacon, Boston,

Durrett, R. (1978). The genealogy of critical branching processes, Stoch. Proc. Appl. **8**, 101-116.

Dynkin, E.B. (1989a). Superprocesses and their linear additive functionals, Trans. Amer. Math. Soc. **314**, 255-282.

Dynkin, E.B. (1989b). Regular transition functions and regular superprocesses, Trans. Amer. Math. Soc. **316**, 623-634.

Dynkin, E.B. (1989c) Three classes of infinite dimensional diffusions, J. Funct. Anal. **86**, 75-110.

Dynkin, E.B. (1989d). Branching particle systems and superprocesses, preprint.

El Karoui, N. and Roelly-Coppoletta, S. (1988). Study of a general class of measure-valued branching processes, a Lévy-Hincin decomposition, preprint.

Ethier, S.N. and Kurtz, T.G.(1986). *Markov processes: characterization and convergence*, Wiley, New York.

Evans, S.N. and Perkins, E.A. (1990). Absolute continuity results for superprocesses with some applications, Trans. Amer. Math. Soc. to appear.

Evans, S.N. and Perkins, E.A. (1989). Measure-valued branching processes conditioned on non-extinction, Israel J. Math., to appear.

Federer, H. (1969). *Geometric Measure Theory*, Springer-Verlag, New York.

Fitzsimmons, P.J. (1988). Construction and regularity of measure-valued branching processes, Israel J. Math. **64**, 337-361.

Fitzsimmons, P.J. (1989). Correction and addendum to Construction and regularity of measure-valued branching processes, Israel J. Math., to appear.

Fleischmann, K. and Prehn, U. (1974). Ein grenzwertsatz für subkritische Verweigungsprozesse mit endlich vielen Typen von Teilchen, Math. Nachr **64**, 357-362.

Fleischmann, K. and Sigmund-Schultze, R. (1977). The reduced critical Galton-Watson processes, Math. Nachr. **74**, 233-241.

Fleischmann,K. and Siegmund-Schultze, R. (1978). An invariance principle for reduced family trees of critical spatially homogeneous branching processes, Serdica Bulgar. math. publ. **4**, 111-134.

Getoor, R.K. (1974). *Markov processes: Ray processes and right processes*, Lecture Notes in Math. **440**, Springer-Verlag, New York.

Gorostiza, L.G. (1981). Limites gaussiennes pour les champs aléatoires ramifiés supercritiques, in *Aspects statistiques et aspects physiques des processes gaussiens*, 385-398, Editions du CNRS, Paris.

Gorostiza, L.G. and Wakolbinger, A. (1989). Persistence criteria for a class of critical branching particle systems in continuous time, preprint.

Gorostiza, L.G., Roelly-Coppoletta, S. and Wakolbinger, A.(1989). Sur la persistance du processus de Dawson-Watanabe stable. L'intervertion de la limite en temps et de la renormalization, preprint.

Halmos, P.R. (1950). *Measure theory*, Van Nostrand, Princeton.

Harris, T.E. (1963). *The Theory of Branching Processes*, Springer-Verlag, New York.

Hoover, D.N. and Perkins, E.A. (1983). Nonstandard construction of the stochastic integral and applications to stochastic differential equations, I,II. Trans. Amer. Math. Soc. **275**, 1-58.

Jagers, P. (1974). Aspects of random measures and point processes. In *Advances in Probability*, P. Ney and S. Port, eds., M. Dekker, 179-238.

Kallenberg, O. (1977a). Stability of critical cluster fields, Math. Nachr. **77**, 7-43.

Kallenberg, O. (1977b). Random measures, 3rd. ed., Akademie-Verlag, Berlin, Academic Press, New York.

Kawazu, K. and Watanabe, S. (1971). Branching processes with immigration and related limit theorems, Th. Prob. Appl. **26**, 36-54.

Knight, F.B. (1981). *Essentials of Brownian motion and diffusion*, Amer. Math. Soc., Providence.

Lee, T.Y. (1989). Conditional limit distributions of critical branching Brownian motions, Preprint.

Le Gall, J.F. (1989a). Marches aléatoires, mouvement brownien et processes de branchement, Lecture Notes in Math. **1372**, 258-274.

Le Gall, J.F. (1989b). Une construction de certains processus de Markov à valeurs mesures, C.R. Acad. Sci. Paris **308**, Série I, 533-538.

Le Gall, J.F. (1989c). Brownian excursions, trees and measure-valued branching processes, preprint.

Le Jan. Y. (1989). Limites projectives de processus de branchement markoviens, preprint.

Liemant, A., Matthes, K. and Wakolbinger, A. (1988). *Equilibrium distributions of branching processes*, Akademie-Verlag, Berlin and

Kluwer Academic Publ., Dordrecht.

Lindvall, T. (1973). Weak convergence of probability measures and random functions in the function space D[0,∞), Journal of Applied Probability **10**, 109-121.

Loeb, P.A. (1979). An introduction to nonstandard analysis and hyperfinite probability theory, *Probabilistic Analysis and Related Topics Vol. 2*, ed. A. Bharucha-Reid, Academic Press, New York.

Luxemburg, W.A.J. (1969). A general theory of monads, in *Applications of Model Theory to Algebra, Analysis and Probability*, ed. W.A.J. Luxemburg, Holt, New York.

Matthes, K., Kerstan, J. and Mecke, J. (1978). Infinitely Divisible Point Processes, Wiley, New York.

Meleard, S. and Roelly-Coppoletta, S. (1989). Discontinuous measure-valued branching processes and generalized stochastic equations, preprint.

Neveu, J. (1986). Arbres et processus de Galton-Watson, Ann. Inst. H. Poincaré **22**, 199-207.

Perkins, E.A. (1988). A space-time property of a class of measure-valued branching diffusions, Trans. Amer. Math. Soc. **305**, 743-795.

Perkins, E.A. (1989). The Hausdorff measure of the closed support of super-Brownian motion, Ann. Inst. H. Poincaré **25**, 205-224.

Perkins, E.A. (1990). Polar sets and multiple points for super-Brownian motion, Ann. Probab., **18**, 453-491.

Prehn, U. (1978) Einige Grenzwerteigenschaften der Lebensbäume kritischer und subkritischer Galton-Watson-Prozesse, Wissenschaftliche Zeitschrift de Pädagogischen Hochschule Dr. Theodor Neubauer Erfurt-Mulhausen, Math. Naturwiss. Reihe **14**, 49-56.

Roelly-Coppoletta, S. and Rouault, A. (1989). Processus de Dawson-Watanabe conditionné par le futur lointain, preprint.

Rogers, C.A. and Taylor, S.J. (1961). Functions continuous and singular with respect to a Hausdorff measure, Mathematika **8**, 1-31.

Sharpe, M.J. (1988). *General theory of Markov processes*, Academic Press, New York.

Ventsel', A.D. (1985). Infinitesimal characteristics of Markov processes in a function space which describes the past, Th. Prob. Appl. **30**, 661-676.

Watanabe, S. (1968). A limit theorem of branching processes and continuous state branching processes, J. Math. Kyoto Univ. **8**, 141-167.

Zähle, U. (1988a). Self-similar random measures I. Notion,

carrying Hausdorff dimension and hyperbolic distribution, Probab. Th. Rel. Fields **80**, 79-100.

Zähle, U. (1988b). The fractal character of localizable measure-valued processes I - random measures on product spaces, Math. Nachr. **136**, 149-155.

Zähle, U. (1988c). The fractal character of localizable measure-valued processes II. Localizable processes and backward trees, Math. Nachr. **137**, 35-48.

Zähle, U. (1988d). The fractal character of localizable measure-valued processes III. Fractal carrying sets of branching diffusions, Math. Nachr. **138**, 293-311.

Author's addresses

Donald A. Dawson
Department of Mathematics and Statistics
Carleton University
Ottawa, Ontario, Canada K1S 5B6

Edwin A. Perkins
Department of Mathematics
University of British Columbia
Vancouver, British Columbia, Canada V6T 1Y4

MEMOIRS of the American Mathematical Society

SUBMISSION. This journal is designed particularly for long research papers (and groups of cognate papers) in pure and applied mathematics. The papers, in general, are longer than those in the TRANSACTIONS of the American Mathematical Society, with which it shares an editorial committee. Mathematical papers intended for publication in the Memoirs should be addressed to one of the editors:

Ordinary differential equations, partial differential equations and applied mathematics to ROGER D. NUSSBAUM, Department of Mathematics, Rutgers University, New Brunswick, NJ 08903

Harmonic analysis, representation theory and Lie theory to AVNER D. ASH, Department of Mathematics, The Ohio State University, 231 West 18th Avenue, Columbus, OH 43210

Abstract analysis to MASAMICHI TAKESAKI, Department of Mathematics, University of California, Los Angeles, CA 90024

Real and harmonic analysis to DAVID JERISON, Department of Mathematics, M.I.T., Rm 2-180, Cambridge, MA 02139

Algebra and algebraic geometry to JUDITH D. SALLY, Department of Mathematics, Northwestern University, Evanston, IL 60208

Geometric topology and general topology to JAMES W. CANNON, Department of Mathematics, Brigham Young University, Provo, UT 84602

Algebraic topology and differential topology to RALPH COHEN, Department of Mathematics, Stanford University, Stanford, CA 94305

Global analysis and differential geometry to JERRY L. KAZDAN, Department of Mathematics, University of Pennsylvania, E1, Philadelphia, PA 19104-6395

Probability and statistics to RICHARD DURRETT, Department of Mathematics, Cornell University, Ithaca, NY 14853-7901

Combinatorics and number theory to CARL POMERANCE, Department of Mathematics, University of Georgia, Athens, GA 30602

Logic, set theory, general topology and universal algebra to JAMES E. BAUMGARTNER, Department of Mathematics, Dartmouth College, Hanover, NH 03755

Algebraic number theory, analytic number theory and modular forms to AUDREY TERRAS, Department of Mathematics, University of California at San Diego, La Jolla, CA 92093

Complex analysis and nonlinear partial differential equations to SUN-YUNG A. CHANG, Department of Mathematics, University of California at Los Angeles, Los Angeles, CA 90024

All other communications to the editors should be addressed to the Managing Editor, DAVID J. SALTMAN, Department of Mathematics, University of Texas at Austin, Austin, TX 78713.

General instructions to authors for

PREPARING REPRODUCTION COPY FOR MEMOIRS

> **For more detailed instructions send for AMS booklet, "A Guide for Authors of Memoirs." Write to Editorial Offices, American Mathematical Society, P.O. Box 6248, Providence, R.I. 02940.**

MEMOIRS are printed by photo-offset from camera copy fully prepared by the author. This means that the finished book will look exactly like the copy submitted. Thus the author will want to use a good quality typewriter with a new, medium-inked black ribbon, and submit clean copy on the appropriate model paper.

Model Paper, provided at no cost by the AMS, is paper marked with blue lines that confine the copy to the appropriate size.

Special Characters may be filled in carefully freehand, using dense black ink, or **INSTANT** ("rub-on") **LETTERING** may be used. These may be available at a local art supply store.

Diagrams may be drawn in black ink either directly on the model sheet, or on a separate sheet and pasted with rubber cement into spaces left for them in the text. Ballpoint pen is not acceptable.

Page Headings (Running Heads) should be centered, in CAPITAL LETTERS (preferably), at the top of the page — just above the blue line and touching it.

LEFT-hand, EVEN-numbered pages should be headed with the AUTHOR'S NAME;

RIGHT-hand, ODD-numbered pages should be headed with the TITLE of the paper (in shortened form if necessary).

Exceptions: PAGE 1 and any other page that carries a display title require NO RUNNING HEADS.

Page Numbers should be at the top of the page, on the same line with the running heads.

LEFT-hand, EVEN numbers — flush with left margin;

RIGHT-hand, ODD numbers — flush with right margin.

Exceptions: PAGE 1 and any other page that carries a display title should have page number, centered below the text, on blue line provided.

FRONT MATTER PAGES should be numbered with Roman numerals (lower case), positioned below text in same manner as described above.

MEMOIRS FORMAT

> **It is suggested that the material be arranged in pages as indicated below.**
> **Note: Starred items (*) are requirements of publication.**

Front Matter (first pages in book, preceding main body of text).

Page i — *Title, *Author's name.

Page iii — Table of contents.

Page iv — *Abstract (at least 1 sentence and at most 300 words).

Key words and phrases, if desired. (A list which covers the content of the paper adequately enough to be useful for an information retrieval system.)

**1991 Mathematics Subject Classification*. This classification represents the primary and secondary subjects of the paper, and the scheme can be found in Annual Subject Indexes of MATHEMATICAL REVIEWS beginnning in 1990.

Page 1 — Preface, introduction, or any other matter not belonging in body of text.

Footnotes: *Received by the editor date.
Support information — grants, credits, etc.

First Page Following Introduction - Chapter Title (dropped 1 inch from top line, and centered). Beginning of Text.

Last Page (at bottom) - Author's affiliation.

[redacted]

[redacted]